Design in America

Selected work by members of the

Industrial Designers Society of America

CONCORD FREE PUBLIC LIBRARY
W9-CVV-287

Design

in

Selected work by members of the

Industrial Designers Society of America

WITHDRAWN

Introduction and text by Ralph Caplan

McGraw-Hill Book Company

New York St. Louis San Francisco

London Sydney Toronto Mexico Panama

745.4
Industrial

Design in America was prepared by the Industrial Designers Society of America, and all of the designs in it were executed by, or under the direction of, its members.

IDSA is a national, non-profit organization established for the purpose of maintaining high standards of professional competence and integrity. It is the only such organization in the United States, and was a founding member of the International Council of Societies of Industrial Design.

The Society consists of some 600 practicing designers and design educators who meet the basic membership requirement: they must have carried major responsibility in mass-market design projects involving the integration of aesthetic, technological and human factors. Such projects include mass-produced objects, packaging, corporate identification programs, product systems, exhibits, space planning, and environmental design. In other words, all the things that make up this book.

Design in America was undertaken in the belief that the clearest and most accurate statement about IDSA and about contemporary American industrial design, resides in the work of IDSA members. Designers submitted work in three general categories: product design, environmental planning and design, and communications design. Each submission was judged by a jury of special competence in its category.

Design credits are carried in the picture captions, with the design firm or corporate staff appearing first, followed by individual designers.

Copyright © 1969 by McGraw-Hill, Inc. All Rights Reserved. Printed in the United States of America. No part of this publication may be reproduced, stored in a retrieval system, or transmitted, in any form or by any means, electronic, mechanical, photocopying, recording, or otherwise, without the prior written permission of the publisher.

Library of Congress Catalog Card Number 75-83268

31685

1234567890 HDBP 754321069

OCT 19 '70

Contents

Introduction: A Divergent Profession

Who do industrial designers think they are?

This collection is a graphic answer to that question. It represents a professional view of effective design done in recent years. The work here was chosen by juries of practicing designers, using criteria somewhat different from those that a critic or layman might use: each entry was evaluated in terms of its designer's own description of the problem he was given and the way in which he solved it.

The result does not reveal a single standard, style, school or trend. Nor, of course, does it represent all areas of American design. The scope of this book is limited to *industrial design, as practiced by members of the nation's professional designers society.*

While that narrows the subject down, it still leaves it uncomfortably broad. For the work of professional industrial designers, as exemplified in this book, includes kitchen appliances and printing presses, perfume packages and traveling exhibits, filling stations and ballpoint pens, trademarks and motion pictures, a system for feeding newborn babies and a system for making office work more efficient and a dredge for controlling river floods during monsoon season. It also includes projects—like space planning and feasibility studies—that may not in themselves consist of a physical design at all.

All of which leads unavoidably to the question, What is industrial design? It was never an easy question to begin with, and it has not, through the years, become any easier. I have never seen a definition that adequately covered all of the projects that industrial design offices undertake. And even if a satisfactory definition were devised, it probably would not cover all of the activities that industrial design offices will undertake a few years from now.

Essentially, industrial design determines the form of objects that are to be mass-produced by machines, rather than crafted by hand. But while this has long been an *essential* definition, it is no longer sufficiently comprehensive, if indeed it ever was. Originally it served to distinguish the industrial designer from the craftsman, who made one-of-a-kind objects. That was a useful distinction, but one that is no longer necessary, adequate, or even valid.

A kind of rough working definition of professional design as currently practiced in the United States can be seen in these pages. First of all, the designs are not necessarily mass produced. The exhibits, the interior schemes, the water storage tank, the motion pictures—these *are* one-of-a-kind products. Nor is the work necessarily done for industry; the Inpost beautification project, the United States Air Force Academy, the UN symbol are all examples of industrial design turned to non-industrial purposes. What most of the selections here do have in common is that they were executed by firms or individual designers committed to a certain kind of practice—a practice that grew out of, and is still most often engaged in, service to industry.

It is, then, the core of the practice, rather than its specific output, that defines and places industrial design. Designers may do any of an astounding variety of things; some come close to a kind of specialty; but all profess to create responsibly within the framework of industrial processes.

As an activity industrial design began when modern industry began. In some respects, its origin was defensive, a protest against the spate of ugly products begat by the industrial revolution. But while this protest was bound to lead (particularily in England) to a wistful attempt to resurrect the craftsman, it also led (particularly in Germany) to an attempt to understand the machine and to develop a machine aesthetic. Thus, in Europe, first in the middle of the 19th century, and later with particular intensity in the twenties, philosophies of industrial design emerged.

They do not seem to have had much effect in the United States at the time. American design began from a theoretical base far simpler than those of the Bauhaus or Werkbund. It was not so much a theory as a suspicion: If products looked better, they would sell better.

The suspicion was felt keenly by a few designers as early as the twenties, and shared by a few clients, who had little to lose. In the prosperous twenties, mass production had loaded the market with competitive products that worked equally well—their distinction, then, presumably had to lie in their appearance. By the depressed thirties, the problem was somewhat different. No one was buying much of anything, unless there was a special reason to. Could design supply that reason?

Apparently it could, and the first industrial designers have been credited with making a major contribution to getting the American economy moving again. While this is no mean achievement, it is not the kind of achievement on which a traditional professional discipline is normally built. As George Nelson observes, "The success of the first designers was based not so much on professional skill or knowledge but on the kind of imaginative common sense one would expect from a first-class salesman."

Some designers, and most clients, thought that was enough. And for certain short-range sales objectives, it may have been. But other designers insisted on more. Confronted with clients who expected a group of quick sketches to respond to, these designers asserted that sketches would be senseless apart from an adequate study of how each product was made, sold, used. This insistence did develop eventually into a philosophy of design (one that corresponded in some respects to that of the Bauhaus), but it was a matter characterized chiefly by either common sense or, in today's jargon, "gut reaction."

The experience of the late Walter Dorwin Teague was a case in point. By the mid-twenties he was a very successful book and advertising designer who, nevertheless, felt "dissatisfied and bored with the work I was doing. Then a few people asked me to do some packages; a piano manufacturer asked me to design two or three grand pianos; a leading automobile manufacturer asked me to sketch a new line of cars with their color schemes. These projects were novel, and I found them stimulating. But none of them worked out very satisfactorily except the grand pianos, which I had built under my own direction. That taught me a lesson."

In 1927, Teague got a chance to apply that lesson, when the Eastman Kodak Company selected him to redesign two of their cameras. He replied that he knew almost nothing about cameras, and would take on the assignment only if he could do the work in the Eastman factory, with Eastman engineers, under a full year's contract calling for one week of work a month.

The client agreed and the association continued until Teague's death in 1960, although Eastman had (with Teague's help) established its own in-house design group long before that.

Teague's experience was not unique. Other of the early American designers report similar experiences; they too perceived almost from the beginning that there could be no responsible design apart from the problems that made responsibility necessary and the information that made responsibility possible. R.H. Macy in 1928 offered young Henry Dreyfuss, who was a stage designer at the time, the job of analyzing merchandise that was not selling well, then making sketches of the forms he thought would be more successful. Dreyfuss refused, having realized that the way to attack the job was to work with the manufacturer at the beginning of a project, rather than with the retailer at the end of one. But, although he turned down the job, the resultant insight led him to become an industrial designer.

In at least that single respect, some industrial designers began their practice with some professional attitudes. They knew, or discovered, that any job was easier once the parameters were known. In the case of product design, constraints clearly had to be drawn from what the product was supposed to do, who used it, which materials and processes were used in making it, and how it was sold. Otherwise, the design might not be practical to make or possible to sell; or the product might betray the consumer through deceptive styling (as many products did, and do).

These constraints would apply even to a project strictly limited to appearance design. But appearance design implied an unrealistic limitation in any case. In order to execute a design that was aesthetically valid, a designer often had to have access to, or develop, precisely the sort of information that could lead to its being made more efficient and more efficiently. And more convenient. Moreover, designing for efficiency and convenience, as well as for aesthetics, was more satisfying to all parties concerned: the designer, the manufacturer, and the consumer.

No industrial designer would regard appearance as *un*important. But appearance might be subordinate to any or all of a number of other design factors: safety, convenience, cost of production, ease of maintenance, wise choice of materials. Appearance might lead a customer to buy, but it could not in itself lead to a satisfied customer. A successful design had to integrate performance with appearance, while appealing to the consumer and making a profit for the manufacturer.

Such design fullness was, of course, an ideal. Rarely did products achieve the perfect blend of function and form. There were too many obstacles. Designers were not all equally gifted; clients were not always sufficiently open minded; neither was necessarily concerned with design integrity; and consumers, thank goodness, did not always reflect the taste of the tastemakers. In addition, the field was too new to have built up a reliable body of expertise. The factory itself was a tool that no one could learn how to use without costly experimentation. As Teague pointed out, "No generation before had had to design by remote control. I might never touch a single one of the objects I designed." The industrial designers of the thirties were apprentices without masters.

But they were learning. They were not all learning the same things; but they did not all have quite the same things to learn. Their backgrounds were too diverse for that. Once the need for an artist in industry was established (or at least expressed), there was no particularly logical place to look for such a person. Searches in the fine arts had been conspicuously unsuccessful, and there was no reason to have expected that they would be otherwise. Industrial designers came from architecture, from commercial graphic design, from theatrical set design, from typography, from fashion illustration, from engineering. Despite the diversity of their origins, it is not really surprising that they developed a common set of principles for, as one of the early designers points out, only those who developed the principles were able to survive.

Those principles turned out to have something in common with the principles of the Bauhaus, although neither the motivation nor the end products were much alike. The Bauhaus had centered on function as the directive force in all design and so, in their own way, had American designers. Europe had a well-hewn theory and some eloquent spokesmen. The United States had almost no theorists or spokesmen practicing industrial design, but it had something more indicative of Yankee value: an economy that could support industrial designers, and that needed them. The game had been invented by Europeans, but the Americans owned the ball. It was natural that the rules conform to what American industry needed. And American industry needed a lot of services that no one other than designers was supplying. The result was that industrial design, once focused upon the object, came to diverge into other paths.

Package design was one such path. Packages for certain kinds of products had long been designed by commercial artists; now that artists were designing products, it seemed logical for the same artists to do the packages as well. Mass marketing patterns were causing American goods to be sold in packages while in Europe the same items were still sold loose or in bulk. The most significant surge of packaging activity came with the introduction, and soon the domination, of the self-service store. As the grocery clerk and his counterparts in other retail markets disappeared, they were replaced, if at all, by the package. At the very least a package had to be quickly identifiable in the confusion of a large supermarket; but more than that, it was a non-vocal (although not always quiet) salesman. Unlike the grocery clerk, however, the package had the job of plugging a particular brand. If a manufacturer's product was going to compete with other brands—in fact, if it was even going to get space on the shelves—it needed a package designed to promote it.

The package, like the product, was a carrier of the company's "public image." There were other graphic materials in this category: letterheads, logotypes, truck panel treatment, uniforms, business cards, brochures, etc. These, like exhibits and interior furnishings, expressed a "corporate identity." The process of designing them in an integrated way, then, was a corporate identity program; and the programmer was an industrial designer.

Many American industrial designers began their training as architects, and one or two maintain successful practices in both architecture and industrial design. While such dual practices are rare, it is not at all rare for industrial designers to engage in certain kinds of architectural endeavors, particularly structures used for selling or promotion: stores, gas stations, exhibit halls, etc. Some offices also perform pre-architectural services, determining the kind and amount of space required for the various tasks to be performed in a building, and thus supplying data for an architect to interpret into form.

Other design offices, directed towards technology, offer design engineering services. Still others provide such management consultant services as market research and product planning. Some of the very large offices are prepared to offer *all* of the above services, as "total design."

How is this possible? It is possible because, as designers began to take on problems so diverse that only a group could have the requisite skills for dealing with them, the "designer" often ceased to be a man and *became* a group. In *Design This Day* (written in 1940 by an industrial designer, but not about industrial design), Walter Teague refers to ". . . this new profession of industrial design, in which one man of restless mind and many interests assembles around him a group of variously trained co-workers—architects, structural and mechanical engineers, painters, sculptors, craftsmen, investigators and writers—and directs their group efforts in an astonishingly wide range of activities." The late Harold Van Doren published his book *Industrial Design* in 1940. By 1954, in a second edition, he felt impelled to add a strong introductory note pointing out that, "Modern product design is almost never the work of one individual." Designer Robert Hose puts it more strongly: "No one person *ever* designs a mass-produced product." The designs of the thirties are almost always credited to individual designers; the designs in this book are usually attributed to firms.

This means of course that an industrial designer rarely has the craftsman's satisfaction of saying, "I made this." On the other hand, few craftsmen can have the satisfaction of controlling the massive and complex projects that industrial design offices take on.

Thus far, I have been talking as if American industrial design were always the product of independent or consultant offices, serving a number of clients in a number of fields. Well, it isn't. Industrial design's earliest practitioners had to be consultants, since, except in the so-called "art industries," manufacturers did not employ designers. But even a cursory glance at the credits in this book will establish that much of today's industrial design is the work of internal designers, employed by manufacturing corporations. Their prominence is not hard to understand. As the need for product design for the mass markets became indisputable, manufacturers found it more efficient to maintain design staffs of their own. Today it is customary for large-volume manufacturers in the consumer goods fields to have their own design staffs, as do many of the specialized high-technology firms.

These corporate designers may occupy management-level positions, particularly in industries where sales are geared to frequent appearance changes. Each of the three major motor companies has a large design or styling division, with a vice-president in charge; in the appliance industry, too, designers serve at several corporate levels.

One of the industrial designers' most-often-cited assets is the general knowledge of materials and fabrication techniques that comes from constant work on a very wide range of products. He presumably brings with it the "fresh outside view," the broad, detached perspective of a mind not involved in the day-to-day details of a corporation's existence. Traditionally, an industrial designer is in a position to understand a firm's problems without being weighted down by them from the inside. How can a corporate designer, however talented and influential, provide the same perspective?

Often he doesn't provide it, and doesn't try to. The closer the corporate designer gets to management, the more he necessarily becomes associated with management views, the more his thinking becomes management thinking. Many firms therefore seek the best of both worlds by maintaining a design staff, and retaining consultants for unusually large projects; for long-range planning, uninhibited by the necessity of keeping up a production flow; for periodic review; for whenever they feel that outside talent can solve problems.

That is the key: solving problems. Whatever else he is or isn't—artist, engineer, salesman, planner, management consultant, inventor—the industrial designer is a problem solver. Some designers go so far as to say that design *is* problem solving (just as others are likely to say that it *is* communication). If taken as definitions such statements are simultaneously too limiting and too pretentious to be helpful. Too limiting, because designers do many things besides solving problems; too pretentious because the practitioners of scores of trades and professions—doctors, plumbers, labor negotiators—also solve problems, most of them problems that no designer should presume to attack.

But such statements are very useful in understanding why what designers actually do goes off in so many different directions: the design-related problems of industry are many and various. The problem may be technological: *How can we isolate the optical elements of a laser from external stresses?* It may be a problem of competition: *How can we modify our product to outsell Brand x?* Or a problem of exploiting resources: *Now that we have invested in these elaborate production facilities, what else can we do with them?* It may be a planning problem: *What should Apex, Inc. be making in 1980?* Or an identification problem: *We're proud of our name—The Indiana Pipe Wrench Company. But, although we have had the name for 50 years, for the last ten we have been making, besides pipe wrenches, toys, architectural components, and scuba equipment. Is there a way of changing our name to correspond to our modern role, without sacrificing the reputation and the recognition we have built up over the years?* It may be a communication problem: *Thousands of prospective customers will attend the Hong Kong Science Fair. Which is the better medium for telling them about our products—an exhibit, or a film?*

The answers to the above questions depend upon more precise definitions of the problem. And that implies an entire battery of further questions. For example, *Film or exhibit?* What do you really want to say to the fairgoers? How much money can be reasonably budgeted for the purpose? Will the design be expected to have any function once the Hong Kong Science Fair is over? (A film can be duplicated and sent anywhere; an exhibit can "travel," but this is a much more cumbersome proposition, and would require special design emphasis.)

Those are of course still only the initial questions. Once some of them are answered, the plans for the Fair itself are studied. How is space allotted? What sort of competition for the visitor's attention will there be? Who is expected to attend, and for how long? What resources does Hong Kong offer for building an exhibit?

And so on. While the project may seem far removed from product design, the design approach is hardly removed from it at all. As in industrial design generally, it consists of defining objectives, determining constraints, and imaginatively devising something that satisfies a large number of people. As a generalist, the industrial designer is often asked to take on problems that fall to him by default, because they are not properly in the domain of anyone's specialty. Frequently he can solve such problems; and the more he does, the more industrial design diverges. For this divergent activity to become a profession in every sense, as its most earnest practitioners wish it to, something more nearly approaching standardized minimal capability is required. That implies educational standards.

Probably no one would contend that present-day design education, as it now stands, is equal to the task. But serious and intense efforts have been and are being made to improve it, with considerable success. The Industrial Designers Society of America has for years been attempting to set and maintain standards.

But the twisting course of design practice in the past has made it difficult for design education to forge a future. Since American industrial design began with a concern for appearances, design training has tended to be concentrated in appearance skills. Thus, many young designers receive training that does not qualify them fully—and may not qualify them at all—for the kind of broad service that industrial design is becoming. Some young designers, on the other hand, receive an orientation too advanced for the kinds of work design offices usually do.

At last count, there were some 2500 undergraduate industrial design students in the United States, with about 500 graduating each year; and some 100 graduate students, half of whom graduate each year. Of the nearly 50 baccaulaureate degree programs in industrial design in the United States, about half offer a Bachelor of Arts or Bachelor of Fine Arts degree. Many of the others include industrial design in schools of architecture or engineering. At least two institutions are accredited to grant their own professional degrees: Bachelor of Industrial Design and Master of Industrial Design. About 30% of all industrial design programs in this country, however, are not taught in colleges or in universities, but in professional art schools. Such schools are sometimes committed to a view of industrial design as "product aesthetics," and to a curricular concentration on rendering and presentation techniques—tools that the industrial designer has always used but that are now rarely his key tools.

IDSA is steadily trying to establish meaningful professional criteria and to find ways of assuring that schools and colleges meet them. Its education committee—made up of both design educators and practicing designers—works to help schools achieve full academic accreditation, and to assure that graduates meet minimal professional standards as well as the individual school's curricular goals. The society's student memberships and merit awards program help keep students and professionals in touch with each other. Probably no single curriculum can accommodate all of the professional tasks an industrial designer may find himself performing. The aim is to establish a reasonable base for them, one that allows for the rapid and sometimes radical change that characterizes design in America.

Fortunately, there is a good deal of support from the students themselves and from recent graduates. Design educator Arthur Pulos, in a report published for UNESCO by the International Council of Societies of Industrial Design, notes an increasing tendency for graduates to take government jobs or to enter such special service programs as the Peace Corps. The most accomplished rendering and styling techniques are not likely to be useful in those positions. Pulos says:

"The practice of industrial design and with it the character of industrial design education has been changing dramatically over the past few years with programs increasing their intellectual content, diminishing their emphasis on product aesthetics and searching for a new position among the established disciplines. A strong shift has been made from design for sales to design for service. . . ."

That shift can be expected to become more highly visible in the next few years as today's students become practicing designers. American design students, like American college students generally, say they are looking very hard for "relevance." Design students have a far better than average chance of finding it; the best of them are uniquely equipped to help solve some of the problems that most need solving—problems in transportation, in food distribution, in urban ecology, in physical medicine.

For years and years industrial designers have been describing themselves as members of an emerging profession. Yet it has never fully emerged; each time it seems about to manifest itself once and for all, another branch appears to make all previous definitions obsolete. Perhaps by now this is inevitable. Perhaps industrial design is destined to be not an emergent, but a divergent profession—one that can never hold still long enough to fit into traditional modes. How to sustain such a profession and how to educate anyone for it are formidable problems. But then, designers are problem solvers.

Ralph Caplan

Part One: Products for People

The Domestic Consumer

The Consumer in Business and Industry

Products for People

We bear in mind that the object being worked on is going to be ridden in, sat upon, looked at, talked into, activated, operated, or in some other way used by people individually or en masse.
—Henry Dreyfuss

Mark Twain once wrote a novel with no weather in it. "Of course weather is necessary to a narrative of human experience," he wrote. "But it ought to be put . . . where it will not interrupt the flow of the narrative." So he included an appendix of rain, wind, dust-storms, and sunshine, and instructed the reader to help himself to the weather of his choice.

Except for references to particular designers, and an occasional model in a publicity photo, there are no people in this book. (Neither is there an appendix of people from which the reader can help himself.) But although the content consists of things and the images of things, implicit on every page is a concern for people—for the men, women, and children who use the things and are affected by the images.

This concern is as good a way as any to distinguish industrial design from design engineering. Each is concerned primarily with a different aspect of product function. The engineer's first responsibility is to see to it that the product works—that the lawn mower cuts, that the vacuum cleaner sucks dust, that the infrared thermometer accurately measures temperatures. The industrial designer's business is to relate that capability to the user.

The lawn mower not only has to cut grass. It also has to *look* as though it will cut grass. And it has to cut grass without also cutting the user's leg or trouser cuff. The vacuum cleaner not only has to suck dust. It also has to be handy to store, to clean, to move.

Those are the product designer's concerns. (They are not his only concerns, nor is he professionally indifferent to the product's prime function, any more than the engineer is indifferent to the people who use his designs.) His contribution lies in his ability to relate people to things.

That relationship may be simple or complex. Design at its most superficial is styling; design at its most substantial is invention. Most of the product design here falls between the two.

Even styling, though, is seldom quite as superficial as it looks and sounds. For product design, no matter how slight, is *three-dimensional* design. Since what the product designer visualizes will be translated by technology into something tangible, certain minimum steps must be taken to establish the market the product is aimed for, the tooling available for making it, the production costs, the competition, the effect of even a slight change on the product's operation.

So a considerable amount of research may be required even for a product treatment that is only skin deep. Some of this may be market research, which some design offices undertake themselves. More likely it is a less formal design-office kind of research. The designer visits showrooms and stores, pretending to be a customer. He may act as a clerk or a gas pump attendant to see what the selling problems are. He will interview the production workers in the factory and may take a turn at operating the machinery. He will certainly collect information on all competitive products. It goes without saying that he will use, and test, and take apart, the existing product, if there is one.

And, as the design begins to take shape, through rough sketches and, perspective drawings, the designer begins to give it genuine shape, in the form of models and mock-ups. Some designers insist that a full-scale model is essential for being sure (and for making sure that the client is sure) of what a product will look like, and also for determining costs.

That much must be done when the problem is one of appearance alone. When designs go beyond that, a myriad of data must be gathered and interpreted. Much of this is technological, and may be supplied by the client's engineers. But a lot of it has to do with "human factors," or what is sometimes called "human engineering." It is obvious that the design of any product ought to take into account the physical attributes and psychological characteristics of the user. How to accomplish this is unfortunately less obvious. The difficulty is compounded by the fact that there is no "user"—rather there may be hundreds of thousands (the designer hopes) of users: male and female, strong and weak, of different heights, having arms and legs of different sizes, eyes and ears of varying sensitivities. Some designers conduct their own human engineering studies. But there are specialists in human engineering, and the designer and his client frequently draw upon their findings and expertise.

Certainly attention to human factors in industrial design is always important, so important that it ought to be taken for granted. Just as certainly, the urgency of human engineering varies with the circumstances. If the dial on a household scale is hard to read, the effect is irritation or perhaps a false sense of dieting success. If the dials in an airplane cockpit are hard to read, the effect may be tragedy.

The Domestic Consumer

The consumer is not a moron;
she is your wife.
—David Ogilvy

Or he is your husband. Or you, whoever you are. *Consumer* is an unattractive term, but it has implications that are useful for our purposes, if not entirely agreeable. It is customary to divide products into categories of "consumer" and "non-consumer" goods. Yet there are no "non-consumers"; there are simply different kinds of consumers. Sometimes they are the same people in different roles. The consumer, for our purposes, is both buyer and user. In domestic life, the buyer-user buys and uses refrigerators and furniture and groceries. At work the buyer-user buys and uses turret lathes and lasers and duplicating machines. At home, the buyer is not necessarily the user. In industry, the buyer is almost never the user.

The range of problems solved by domestic designs is extraordinarily wide. They change as living habits change, and the design solutions may themselves change living habits. In many parts of the United States, room air conditioners have ceased to be luxury objects and have become essential amenities. The design problems they now raise have less to do with improving operation than with diminishing presence. The room air conditioner on page 18 is designed to be as unobtrusive as possible, ideally neither seen nor heard.

The Eames lounge chair and ottoman (page 14), on the other hand, is designed both to be sat in and to be seen. It has a presence, derived in part from an unabashed opulence, that makes it impossible to ignore, or to want to ignore.

A fairly common product design aim is

to express a technological advance and, in the process, enhance it. The speaker systems (page 38, 39) do this, as does the Lytegem (page 16), which takes the principle of the high intensity lamp—extreme concentration of low wattage—and incorporates it into a versatile and pleasing form. The amplifier on page 41 is intended to make a strong statement of professional quality. A minor but interesting means to this end is the use of large clearance holes that let the knobs project through the front panel, suggesting that the important functions occur *inside* the unit.

A design constant in almost every home and office is the telephone (page 35), and there are many features peculiar to it as a design problem. For one thing, with few exceptions, telephones are not sold to the user. You do not buy your telephone, you subscribe to a service; and the instrument is the major component of the service system. More nearly ubiquitous than any other single designed product in daily use, the telephone must be operable by virtually everyone and compatible with any interior. Its use ranges from the trivial to matters of life and death. The ongoing development of telephone handsets represents thousands of man hours on the part of designers and engineers.

A designer, while trying to make product use easy, may try to make product misuse difficult. The socket wrenches on page 47 have a handle taper that simultaneously makes them easier to grip firmly but very awkward to hammer with. This turns out to be important, since the product is unconditionally guaranteed, regardless of abuse by the user. In the guarantee, no holds are barred; but in the manufacture, one hold is at least discouraged by design.

That points up the fact that it is not necessarily the biggest or most glamorous products that inspire design attention. Nothing is less awe-inspiring than a shower ring, yet it is not always easy to find good ones. The one on page 16 has a simple, effective locking device, and is unassertive enough to go with any curtain design or any bathroom decor.

Another humble product with an extremely thoughtful and interesting design is the 6-quart utility bucket on page 45. It is extremely versatile, doubling as a paint roller pan when a screen is added. The injection-molded polyethelyne sheds paint, forms a seam-free interior. The flat surface permits the bucket to be used conveniently on stairs, or on a ladder; the surface is also handy for scraping, or for getting excess paint off the brush. The handle, which will hold a brush or rag or sponge, is notched to keep the bucket perfectly level when it is suspended. The bottom of the bucket is sloped towards the front, creating a finger ledge that gives a firm grip for pouring.

The Cutco knives on page 26 represent one designer's unique and massive research brought to bear upon a universal problem: extending and accommodating the human hand through handles. In developing the Wedge-Lock handle, Thomas Lamb spent several years in learning the intricacies of hands. One of the results is the superior set of handles shown here. They are safe and efficient to grip; they fit hands of all sizes, left and right; and they can be used without fatigue.

Occasionally a design problem is not stated in terms of a specific product, but in terms of a company objective. The objective of the Foamold Corporation was to discover new uses for its product: bulk polystyrene expanded bead foam. The designers first determined that existing uses would be greatly expanded if there were a handier way of shaping the foam, then developed the light-weight, easily operated Foam Shaper on page 47. Ideal for the preparation of custom-fit protective packing for delicate instruments, the instrument makes it possible for anyone to cut unusual shapes precisely; and the Nichrome cutting wire can be pre-formed to cut great lengths of intricately shaped extrusion-like foam. The product has both professional and household uses.

Lounge chair with ottoman

Designer: Charles Eames
Client: Herman Miller Inc.

Molded plywood sections on swivel base, snap-on leather cushions contain down-filled envelopes of foam rubber.

Polaroid 100 Camera

Designers: Henry Dreyfuss Associates
Client: Polaroid Corp.

Design is intended to support the qualities of the engineering: convenience, sturdiness, simplicity.

Silver serving pots

Designer: L. Garth Huxtable
Client: Restaurant Associates, Inc.

Part of the more than 100 service items designed for the Four Seasons Restaurant in New York.

High-intensity lamp

Designers: Michael Lax & Associates: Michael Lax
Client: Lightolier, Inc.

Geometric elements of shade and base selected for balanced relationship in various positions. Shade constructed of internal reflector and spherical shell, with air space in between for cooling, connected with black plastic glare band. Molded plastic base; chromeplated telescoping arm.

Shower ring

Designers: Francis Blod Design Associates, Inc.: George Stehl
Client: Joseph Kaplan & Sons

Flexible Butyrate ring is forcibly sprung over tubular curtain rod, then fixed with molded snap catch.

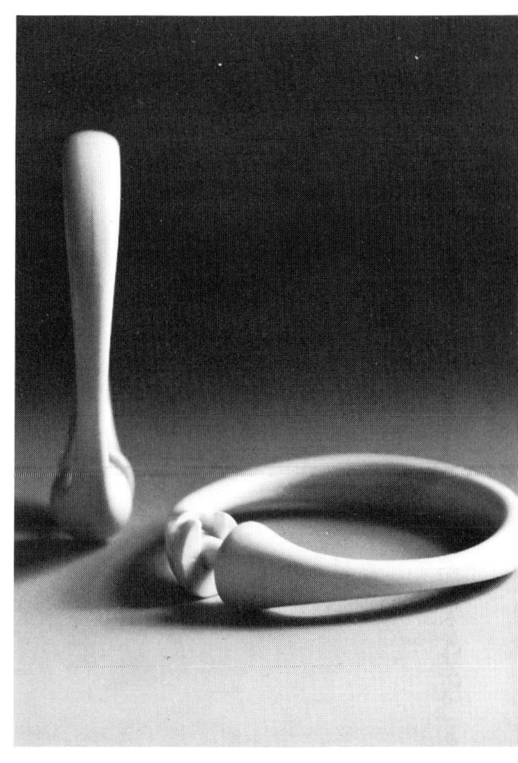

Electric blanket control

Designers: Walter Dorwin Teague Associates, Incorporated
Client: Fieldcrest Mills

Gold band makes decorative feature of joint between upper and lower Melamine housings. Numerals are concealed when blanket is turned off, to minimize "instrument" look.

"Water Pik" oral hygiene appliance

Designers: Raymond Loewy/William Snaith, Inc.
Client: Aqua-Tec Corp. Div. of Teledyne

Inverted case holds water, which is pulsed through "pic" with adjustable force. When interchangeable pic is removed, tube and hand grip recede into casing. Plastic housing is white and dove gray.

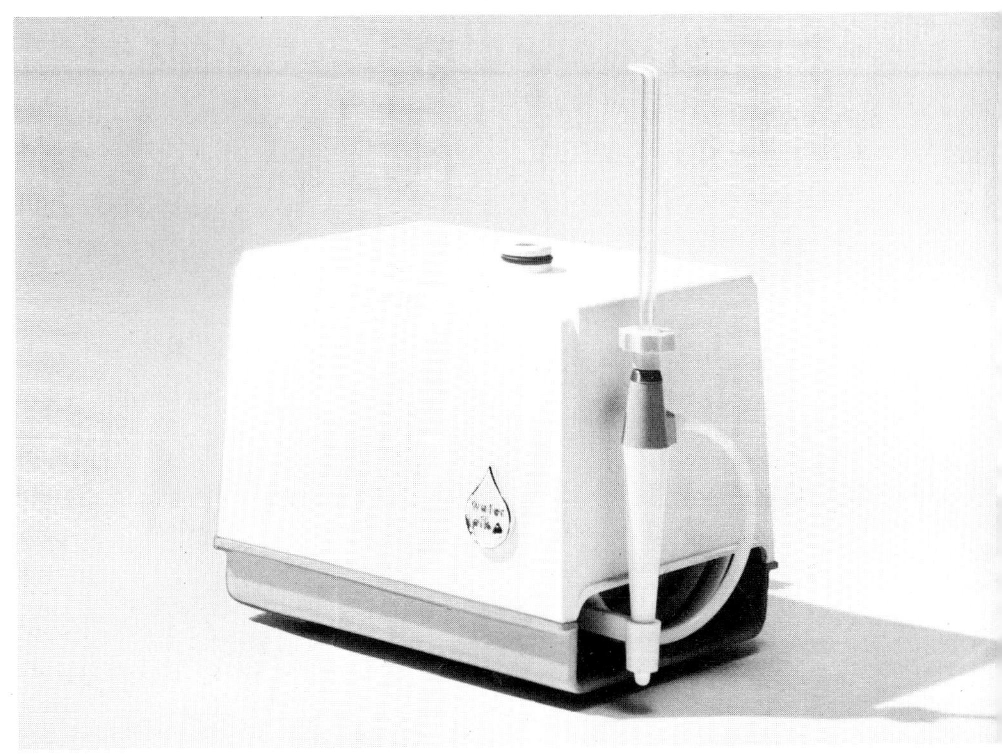

Refrigerator/freezer

Designers: Westinghouse Electric Corp. staff: R. W. Kennedy, W. H. Appel, C. F. Graser

French doors and freezer drawer cut down space-consuming door sweep, help solve right-or-left-opening problem. Design solution accented by full-length handles. Interior arrangement is flexible, with adjustable shelves in cabinet and in door.

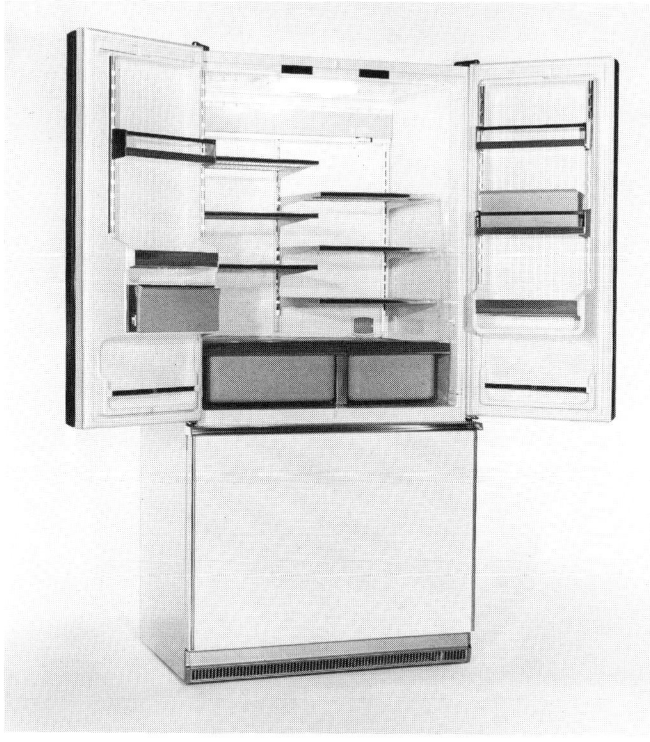

Room air-conditioner

Designers: Westinghouse Electric Corp. staff: R. M. Stoughton, W. H. Appel, C. F. Graser

Panel front acts as noise-reduction baffle. When lower panel is raised, controls are exposed and upper panel drops. Flush panels when machine is not in operation permit draperies to be drawn without interference.

Gas furnace

Designers: Waltman Design: Charles T. Waltman
Client: Lennox Industries

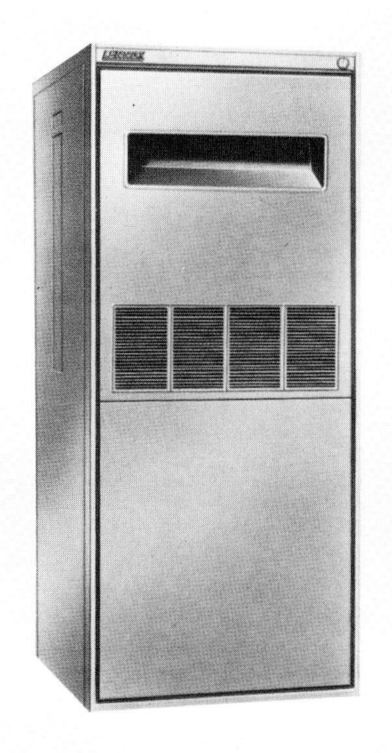

Refrigerator freezer with decorative panels

Designers: Philco-Ford Corp. Staff: Alphonse M. Marra; R. E. Munz, vice president

Problem was to design easily applied decorative panels to go with custom designed kitchen cabinets. Fully extruded door frame has one side or all parts removable. Bail handle is placed on wide extruded band to provide for flat panels with no holes necessary.

Electric hot plate

Designers: Richardson/Smith Inc.: Terry J. Simpkins, Jack L. Petrick, Deane W. Richardson, David B. Smith
Client: Trak Inc.

Manufacturer's aim was to upgrade the status of the hot-plate and expand its market. The compact respectability of the redesigned product makes it an auxiliary appliance acceptable for buffet use, for snacks, for outdoor cooking.

Beverage warmer

Designers: Corning Glass Works staff: T. J. Ryan

Keeps hot beverages at serving temperature of 187°. Compact (7" × 12") unit accommodates two standard decanters, can be built into counter. Control area is tilted for legibility and to contain spilled liquids. Heating elements are printed circuits sandwiched between layers of glass-ceramic. Unit is beige, black and white.

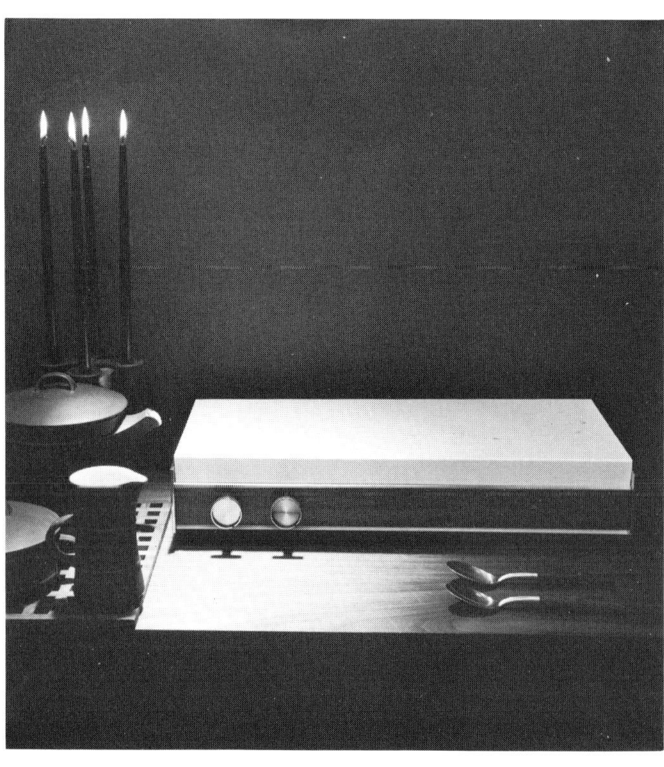

4-slice toaster

Designers: Housewares Division, General Electric Co. staff: M. C. Hauenstein, O. E. Haggstrom, Manager, I.D.

Chrome-plated steel shell with black plastic ends. Handle surfaces are recessed to allow application of decorative identifying panels for subsequent model changes. Hinged crumb tray on bottom facilitates cleaning.

"Toast-R-Oven"

Designer: Housewares Division, General Electric Co. staff: R. Funk, P. O. Rawson, O. E. Haggstrom, Manager, I.D.

Toasts, bakes, top-browns; also reheats foods.

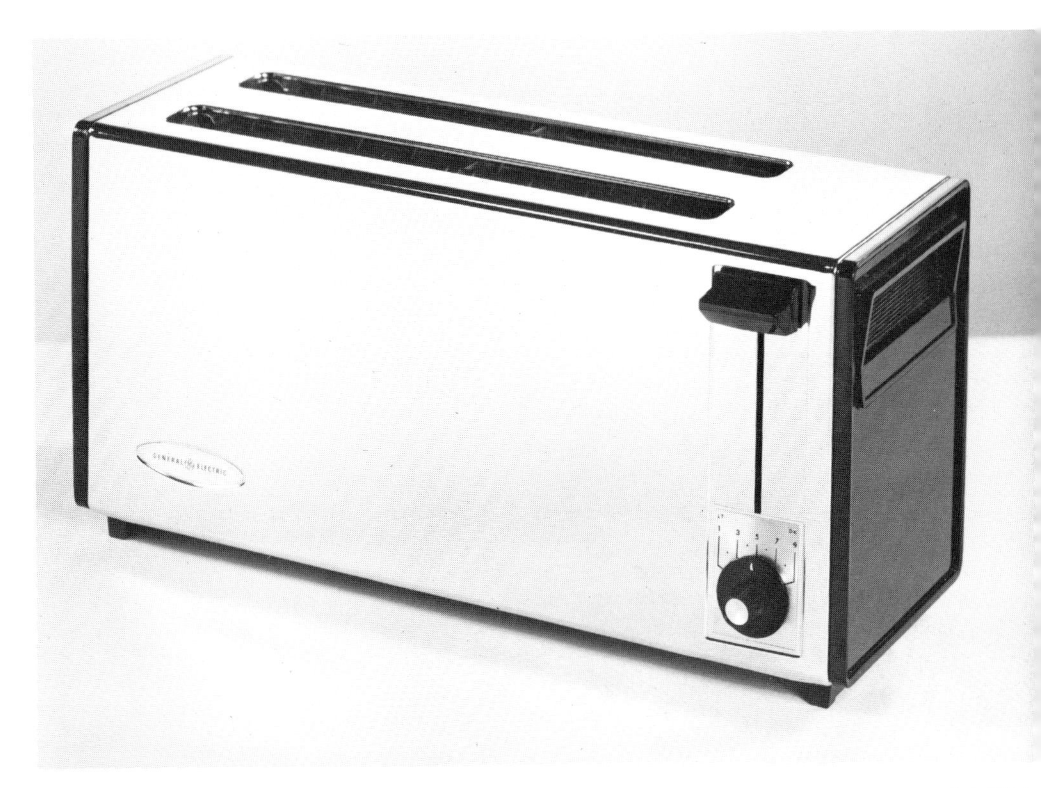

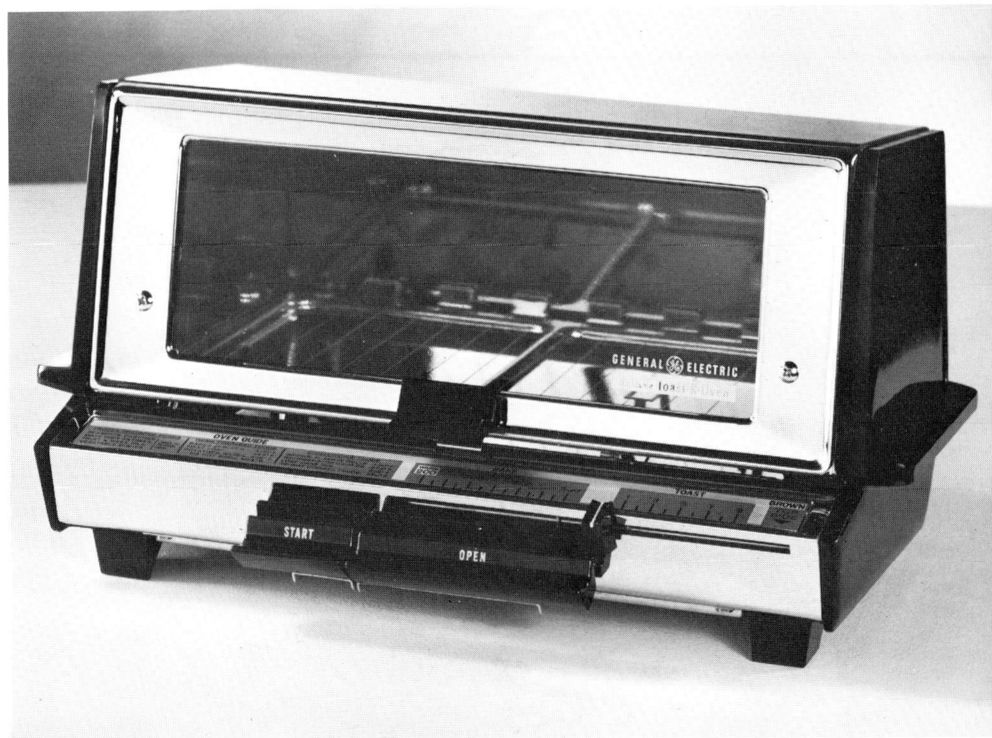

Mixer

Designers: Dave Chapman, Goldsmith & Yamasaki, Inc.: Robert LeSueur, Marlan Polhemus, Kim Yamasaki
Client: Hamilton Beach Co.

Rigid lower housing fits perfectly into semi-flexible upper motor chassis.

Instant coffee/tea brewer

Designers: Mel Boldt and Assocates, Inc.
Client: National Presto Industries, Inc.

Device for brewing two-to-five cups of instant coffee or tea has Lexan body and lid, walnut handle.

Blender

Designers: Dave Chapman, Goldsmith & Yamasaki, Inc.: Kim Yamasaki, Robert LeSueur
Client: Hamilton Beach Division of Scovill Mfg. Co.

Chrome and gray plastic housing, clear plastic container. Seven-speed control; "Auto-Spatula" guides ingredients into blades.

"Air insulated" percolator

Designer: Alfred W. Madl
Client: John Oster Manufacturing Co.

Automatic electric percolator keeps coffee hot for duration of meal without electricity. Except for heating element, pump, and basket, the percolator is entirely plastic: plug tunnel of melamine, inner base of phenolic, body shell and spout molded of polycarbonate for heat resistance, impact strength, dimensional stability. An air pocket between walls serves as the insulator.

Blender

Designers: General Electric Co., Housewares Div. Staff: W. J. Cook, P. O. Rawson, O. E. Haggstrom, manager, I. D.

Compactness, a major objective, was achieved by housing the motor behind the jar and using a notched belt to drive the jar spindle. Vinyl funnel lid with clear plastic cap insert allows addition of ingredients during blending.

Electric knife

Designers: Dave Chapman, Goldsmith & Yamasaki, Inc.: Robert LeSueur, Marlan Polhemus, Kim Yamasaki
Client: Hamilton Beach Div. of Scovill Mfg. Co.

Existing electric knives were awkward to use because motor and housing were too large to grasp comfortably. By moving the motor off the direct axis of blades, lowering it, and angling it, designers made handle an extension of blade.

Food grinder and chopper

Designers: Downer P. Dykes Industrial Design: Downer P. Dykes
Client: Rival Manufacturing Co.

Contemporary adaptation of old-fashioned clamp-on device incorporates food transfer and cutter mechanism used in similar appliances made by client. Aluminum food hopper assembly is easily detached for washing.

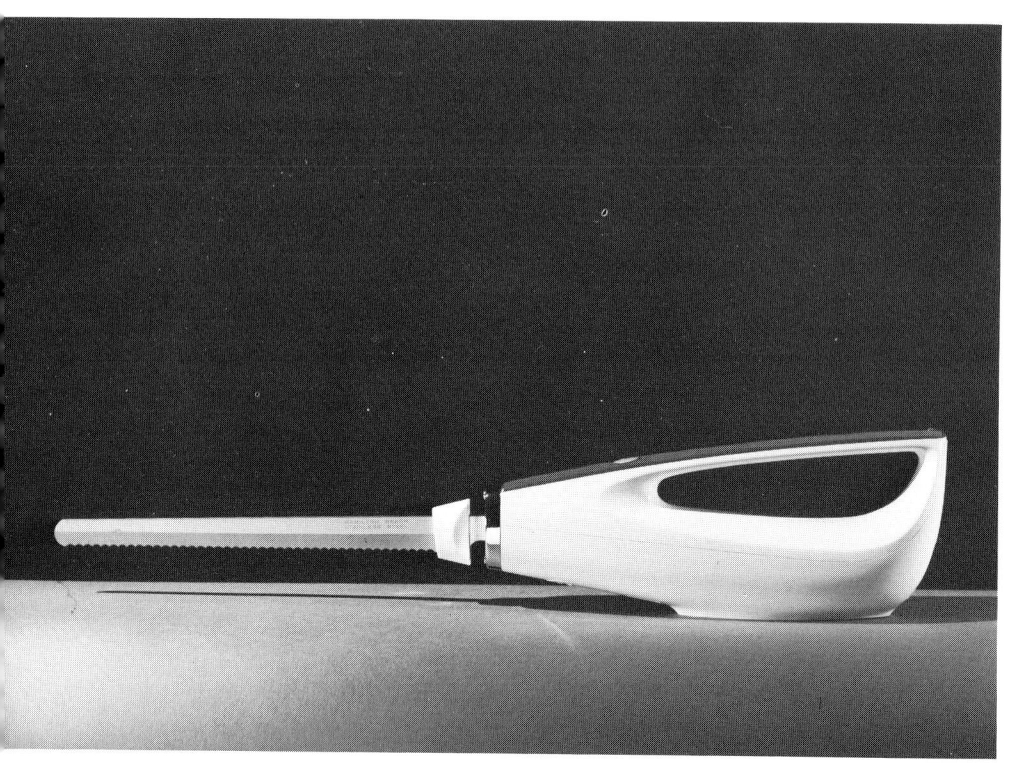

Cast iron cookware

Designers: Michael Lax & Associates: Michael Lax
Client: COPCO, Inc.

Cast iron cookware with porcelain enamel finish is designed for stove-to-table service. Integral cast handles on casseroles afford production economies, ease of carrying and storage. To reduce weight, forms are designed for thin wall-casting techniques and covers made of stamped sheet steel. Bright colored exteriors, white inside. Logo and packaging designed as part of same program.

Egg beater

Designers: Latham Tyler Jensen, Inc.
Client: Ekco Housewares Co.

Assembly is radically simplified by gearing arrangement based on large gear wheel with teeth on outside. Special fastener was devised to attach knob to rotational handle. Die-cast frame with molded plastic handle and knob.

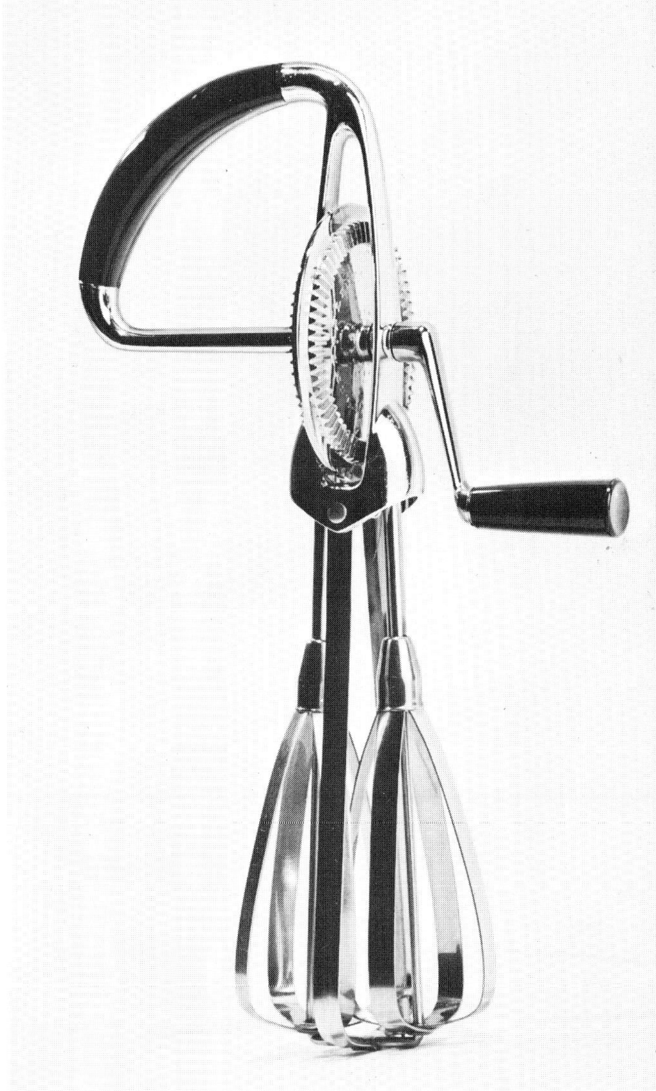

Cutlery

Designers: Michael Lax & Associates: Michael Lax
Client: Ontario Knife Co.

Inventory problems and tooling costs were reduced by minimizing number of knives; yet this line meets all domestic cutting needs. Flat ground, hollow ground or serrated blades, depending on function, all have molded Nylon handles.

Cutlery Handles

Designer: Thomas Lamb
Client: Cutco division of Wearever Aluminum

Based upon years of research into the mechanics of the human hand, these handles provide maximum comfort and efficiency, irrespective of the user's hand size.

Spatula

Designers: Raymond Spilman Industrial Design: Rod Lopez-Fabrega, Raymond Spilman
Client: Androck, Inc.

Versatile kitchen tool has paper-thin blade sculptured for rigidity in scraping or lifting, yet flexible for serving pie, turning eggs, flipping pancakes, icing cakes. Hole in blade can be used to drain excess grease or as egg separator. PB-1-A

Cutlery

Designers: Latham Tyler Jensen, Inc.
Client: Ekco Housewares Co.

Forged satin stainless steel is taper-ground. Handles of carved impregnated Pakkawood.

Kitchen tools

Designers: Latham Tyler Jensen, Inc.
Client: Ekco Housewares Co.

Stainless steel kitchen tools have Pakkawood handles, rivetless union of working parts to stalk, full-tang construction.

Skillet

Designers: Peter Muller-Munk Associates, Inc.: Glenn W. Monigle
Client: Griswold Manufacturing Co., Div. of the Randall Co.

The elimination of almost all handwork substantially lowered finishing costs. Raising handle boss above body proper allows automatic finishing of body exterior surfaces. Accent-colored porcelain enameled steel lid replaced a more costly cast-aluminum lid and, in conjunction with anti-burn shield under the lid knob, provides "high-style."

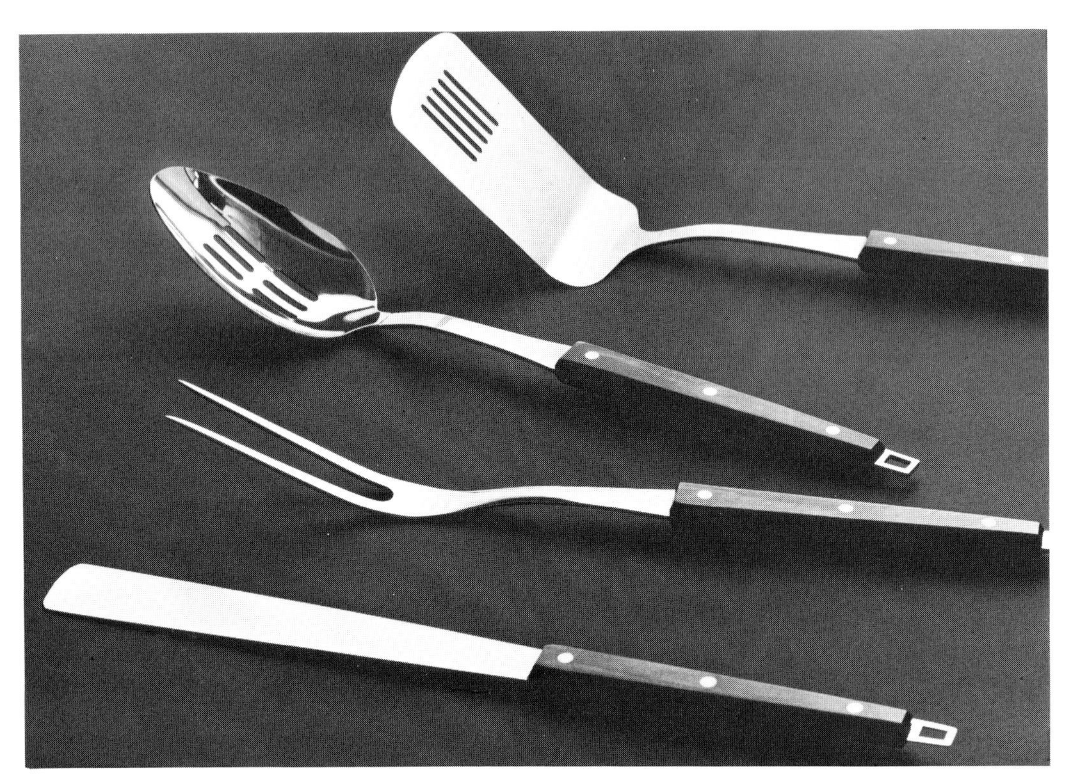

"Tap-Icer" ice cracker

Designers: Donald Deskey Associates, Inc.
Client: The Frank Earnest Co.

Product formerly had steel shaft, which designer replaced with polypropylene, reducing fabrication costs and improving utility. When shaft is removed from mold, two stainless steel heads are snapped into cavity.

Glass-ceramic mugs

Designer: Thomas J. Ryan
Client: Corning Glass Works

Eleven-ounce mugs for hot or cold beverages are molded in one piece of Pyroceram. Three-finger-grip handle has flat top for thumb.

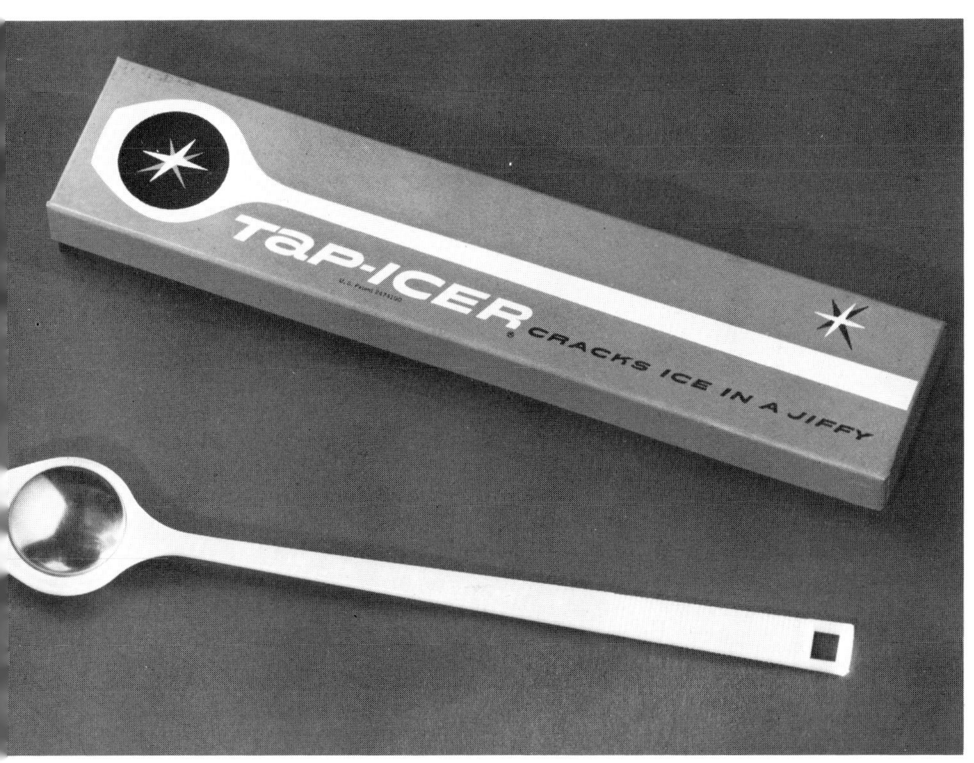

Glassware

Designers: Latham Tyler Jensen, Inc.: George B. Jensen
Client: Rosenthal Porzellan A. G.

Geometrically formal effect of perfectly cylindrical bowl and square pedestal is softened somewhat by curving stem.

Candle and holder

Designers: Raymond Loewy/William Snaith, Inc.
Client: Joseph P. Kennedy, Jr. Foundation

Developed to be produced by mentally-retarded persons and sold in department stores to help the handicapped become self-supporting.

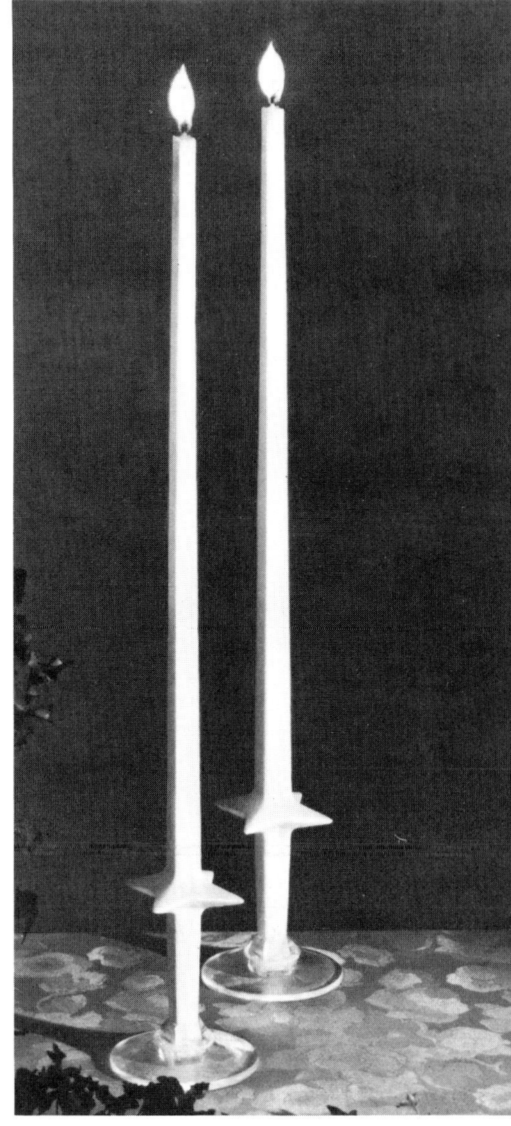

Glassware

Designers: Latham Tyler Jensen, Inc.: Richard S. Latham
Client: Rosenthal Porzellan A. G.
All pieces in this set of formal glassware are the same height, with bowl depth varying as required.

Cook-and-serve ware

Designers: Latham Tyler Jensen, Inc: Richard S. Latham
Client: Rosenthal Porzellan A. G.
High-fire porcelain suitable for oven and range-top high-temperature applications, and also for refrigerator storage.

Sling sofa

Designers: George Nelson & Co., Inc.
Client: Herman Miller Inc.

Problem: Design a sofa system with more comfortable suspension. Solution: make a loop; add intermediate pieces with leg attachments as required for two, three, and four-seater units. Use rubber membrane and rubber straps for suspension, with loose leather pillows of black or tan.

Table

Designers: Howell Design Corp.: James A. Howell
Client: Tri-Mark Designs

Glass-topped table with chrome-plated steel base has two curved leg members half-lapped to interlock with each other. Can be knocked down by removing two recessed socket-head screws under joints in legs. Clear or gray glass.

Chairs

Designer: Charles Eames
Client: Herman Miller Inc.

Side members and swivel base of cast aluminum. Naugahyde upholstery is electronically welded, held in tension.

Outdoor furniture

Designers: Design West Inc.
Client: Samsonite Corp.

Weather resistant medium-priced outdoor seating has extruded ABS slats interspaced with injection-molded ABS beads and assembled on plated steel rod, which provides seat contours and semi-spring support. Steel tubing frame is designed to stack. Protective coating provided by zinc phosphatized base, flow coat primer, baked enamel top coat.

Bubble lamps

Designers: George Nelson & Co., Inc.
Client: Howard Miller Inc.

Design based on Navy's method of mothballing ships: a self-webbing plastic sprayed over objects becomes a protective skin. Metal wire cage is cocooned with warm white translucent skin for softening incandescent light sources.

Telephone set

Developed by Bell Telephone Laboratories–industrial design collaboration with Henry Dreyfuss Associates

"Touch-Tone Trimline" has illuminated dial mounted in handset, enhancing ease of use and permitting placement of phone in areas that might not accommodate conventional sets. Bases come in table and wall models.

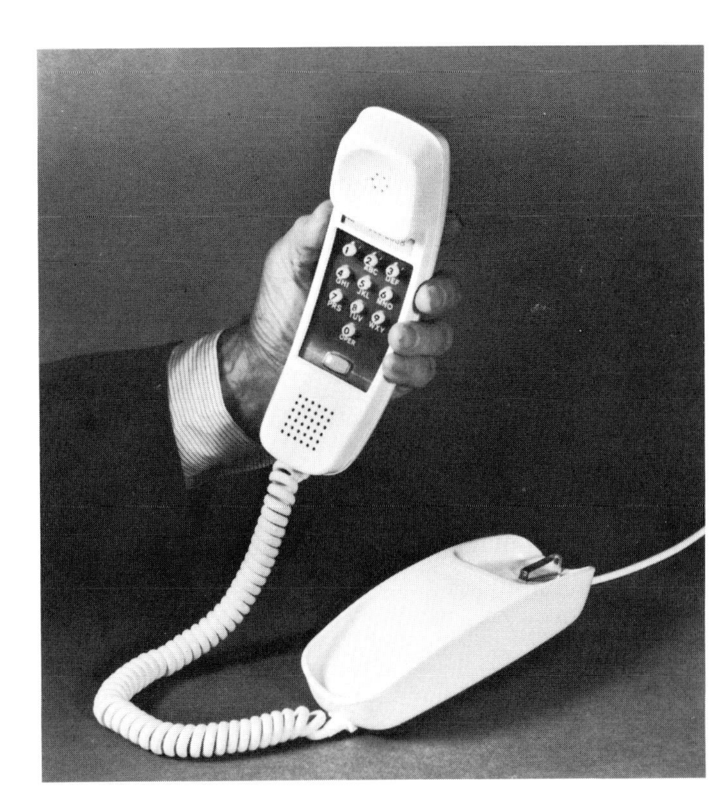

Telephone set

Developed by Bell Telephone Laboratories–industrial design collaboration with Henry Dreyfuss Associates

One of the first uses of a plastic for a mass-produced item–one that has to fit a great variety of locations and of hand and head sizes.

AM-FM clock radio

Designers: Dave Chapman, Goldsmith & Yamasaki, Inc.: Kim Yamasaki, Marlan Polhemus, Ron Guttenberg. Client's staff: Tucker P. Madawick, B. A. Grae

Client: RCA *Sales Corp.,* Consumer Electronics Division

Compact unit has top speaker, with grille fabric integrated into general exterior styling.

Stereo tape cartridge

Designers: Alfred Wakeman Industrial Design: Alfred W. Wakeman

Client: Audio Devices, Inc.

Designed for automobile use, the cartridge is sold to recording companies for loading under various brand names.

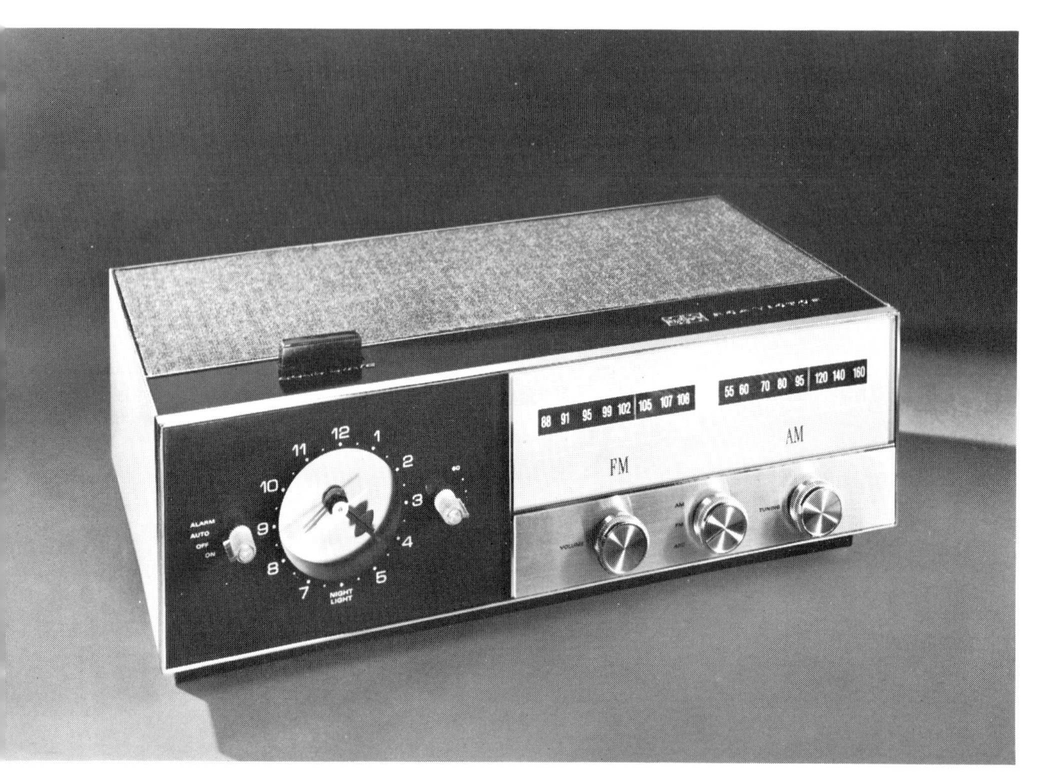

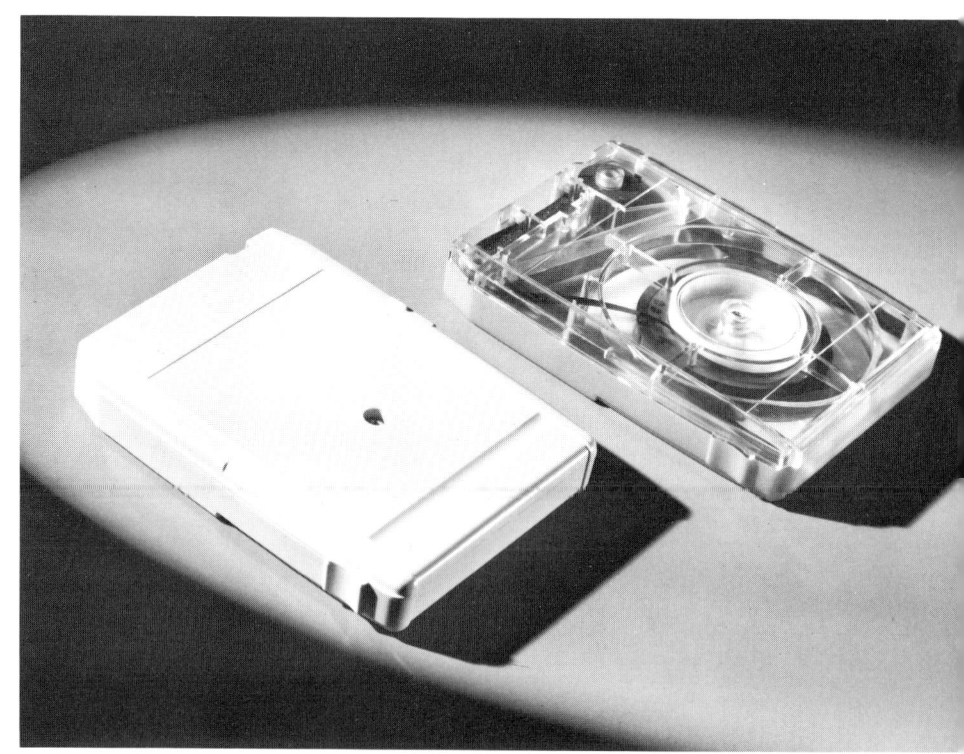

AM-FM-AFC portable radio

Designers: Philco-Ford Corp. staff: James Dermody, Richard Greenwood, designers; R. E. Munz, vice-president

Painted, brushed, and perforated front treatment.

Stereo energizer

Designers: Arnold Wolf Associates: Arnold Wolf. Client's staff: Lamont J. Seitz, Project Director

Client: James B. Lansing Sound, Inc.

Free-standing variation of factory-installed unit; displays internal equalizer network information through dark gray Plexiglas window. Pilot-light spill is reflected off gold-flashed printed circuit board.

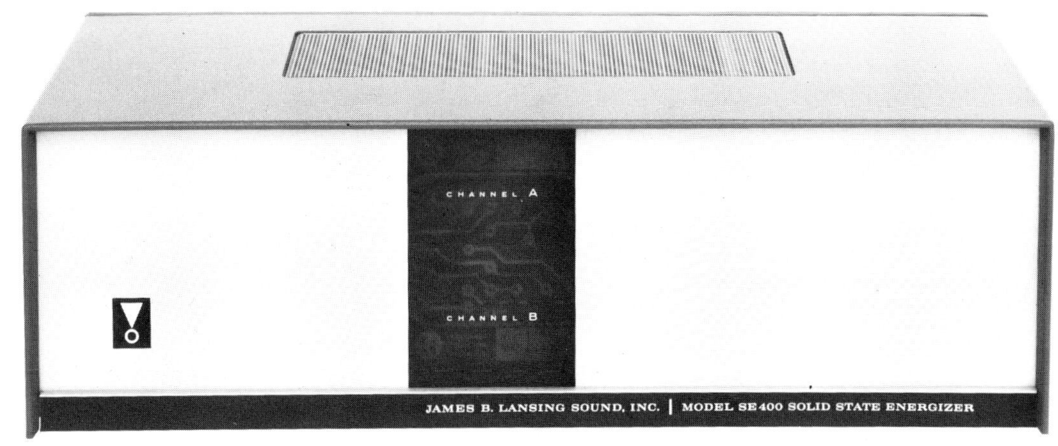

Stereo cassette player and AM-FM tuner

Designers: Philco-Ford Corp. staff: James Dermody, designer; R. E. Munz, vice-president

Controls arranged horizontally, tuning dial runs vertically. Pop-up access to cassette.

Speaker system

Designers: Arnold Wolf Associates: Robert Onodera, Arnold Wolf
Client: James B. Lansing Sound, Inc.

Design is directed particularly to "young marrieds." Behind the round grille is a 12" woofer; rectangular grille screens a high-frequency unit and an open port. Walnut surface is carried around to front face.

Hi-fi components

Designers: Latham Tyler Jensen, Inc. Denmark: Jakob Jensen, Managing Director
Client: Bang & Olufsen A/S

Variable functions controlled by positive-action Vernier slide rules; all other switching by push button. Speaker is solid cube of wood milled out to receive six tweeters with perforated aluminum covers, produces "non-directional stereo sound," said to be identical in all parts of listening room. Amplifier and tuner escutcheon panels of anodized extruded aluminum, screw-mounted to cabinets of rosewood, teak or light oak.

Portable phonograph

Designers: RCA Staff Design: B. Grae, Mgr., Product Design; T. P. Madawick, VP, Industrial Design

"Swingline" cabinet narrows silhouette when closed, becomes "see-through" cabinet when open. Remote switch controls the counter-balanced automatic turntable, which floats down into position as removable speaker wings swing open.

Solid-state amplifier

Designers: Arnold Wolf Associates: Arnold Wolf. Client's staff: Lamont J. Seitz, Project Director
Client: James B. Lansing Sound, Inc.

Most-frequently used control knob (volume) occupies isolated central position; the next two most frequently used are located at outer extremes. The front panel's two functional domains are differentiated by use of vertical and horizontal metal brushing. New lever switch handles use available rocker switch mechanisms; knobs are produced from solid extrusion.

35 mm slide projector

Designers: Michael Lax & Associates: Michael Lax
Client: Airequipt, Inc.

Cycolac housing contributes to lighter weight, increased durability, quiet operation, and lower manufacturing costs.

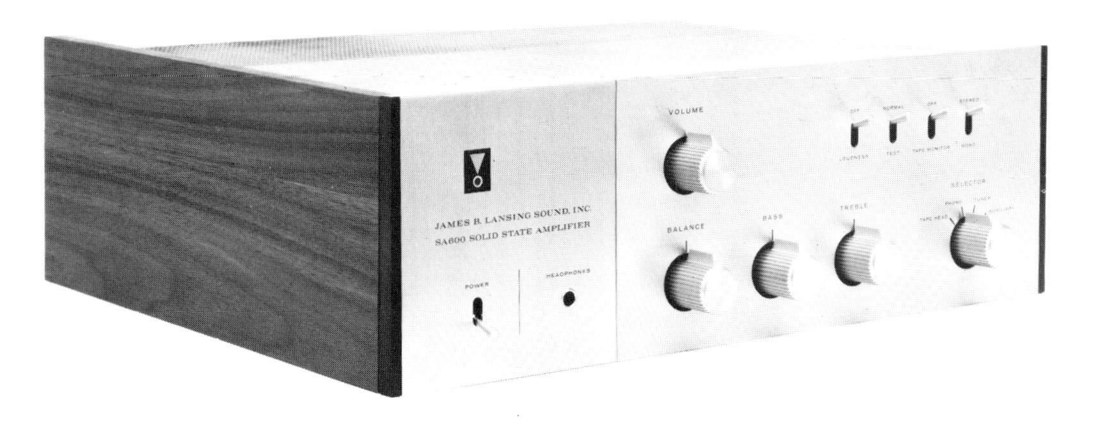

Portable tape recorders

Designers: Herbst-LaZar Industrial Design: W. Herbst, R. LaZar, R. Bell
Client: Artic Import Co.

Three items in a line of low-to-moderately priced battery-operated recorders.

Portable television set

Designers: RCA Staff Design: B. Grae, T. P. Madawick, VP, Industrial Design; in collaboration with Dave Chapman, Goldsmith & Yamasaki, Inc.: M. Polhemus

Spring-loaded control deck remains out of sight until activated by push button. Adjustable leg system allows choice of viewing angles.

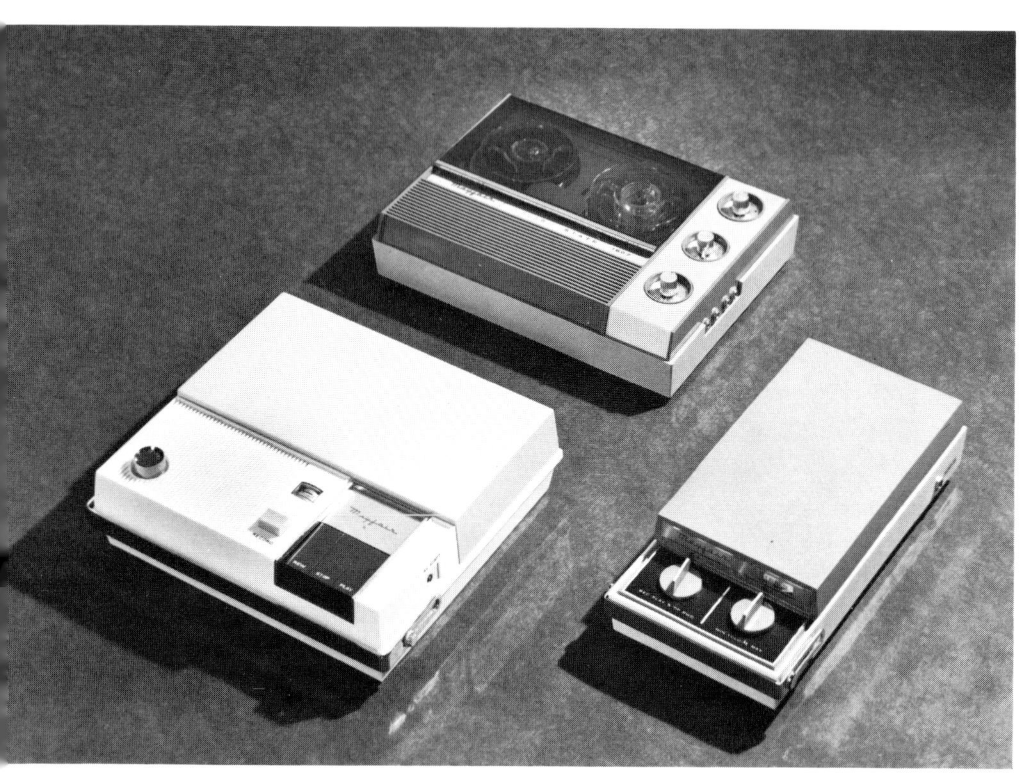

Portable intercom

Designers: Yang/Gardner Associates, Inc.: Peter Quay Yang
Client: Fanon Electronic Industries

Talk-Listen push bar is locked by sliding name plate forward. Master and remote unit use same molded plastic bottom housing. Top cover is changed as model variations require.

Desk Clock

Designers: Dave Chapman, Goldsmith & Yamasaki, Inc.: Douglas Anderson, Kim Yamasaki
Client: Jefferson Electric Co.

Die-cast housing with phenolic base (cord reel in base). High-polish chrome finish and black face are meant to appeal to masculine taste.

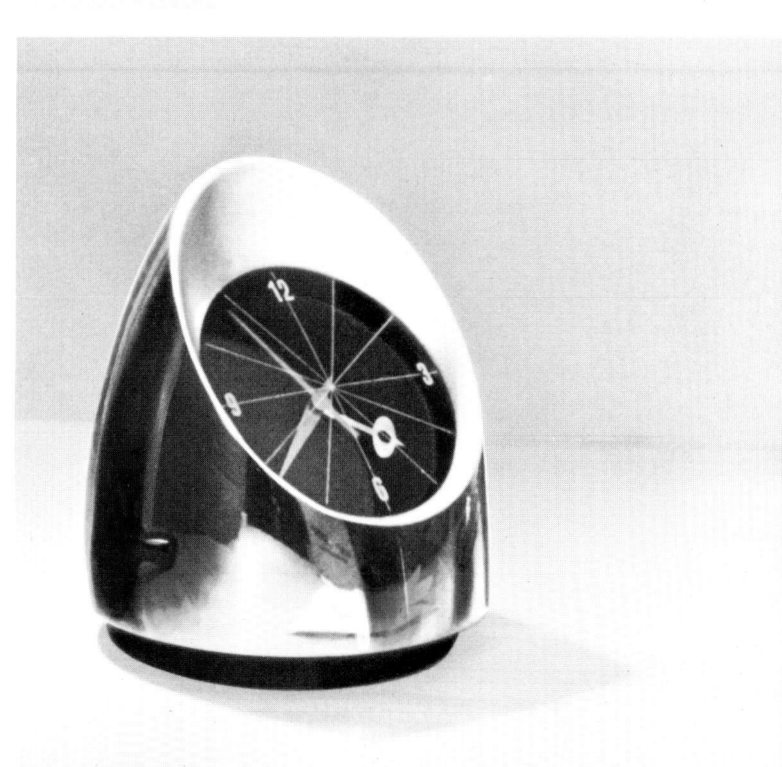

"Estate Keeper"

Designers: Brooks Stevens Associates: Dave Nutting, Darrel Lauer
Client: Bolens Division of FMC

Powered by 6-HP engine, has 6 different lawn-tending attachments mounted forward of front wheels.

Steelmaster measuring tapes

Designers: Laird Covey Industrial Design: Laird Covey
Client: Stanley Tools division of The Stanley Works

Shape permits compact tape rolling, and allows corner fastening of case halves. When folded, crank assembly is visually integrated with hub. Yellow Mylar label matches color of rule. Canoe-shaped markings on rule are easy to spot.

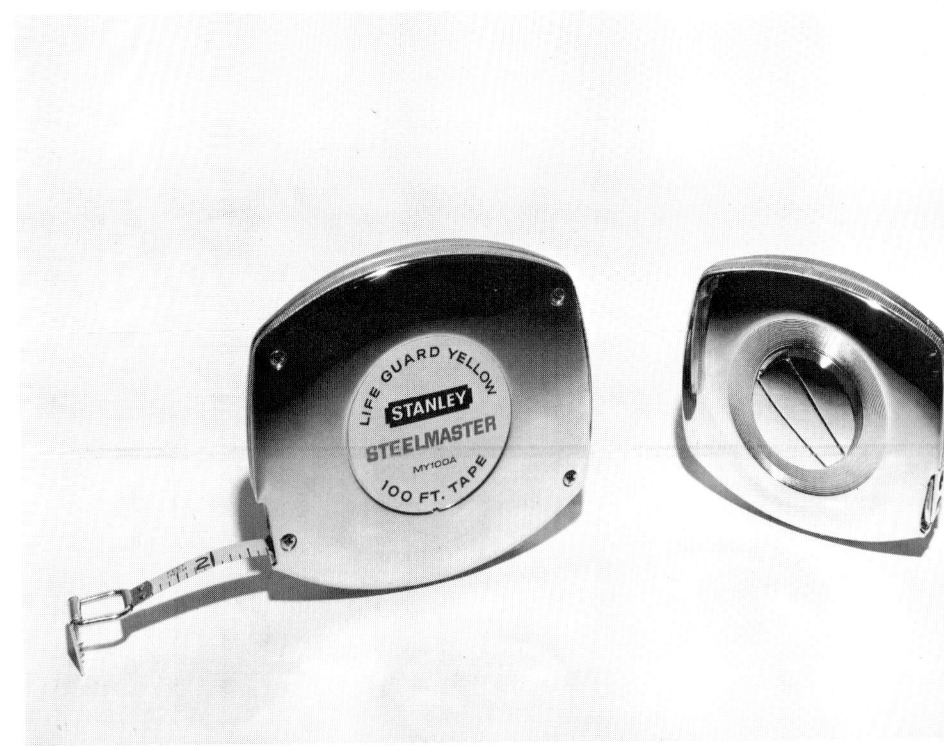

Household utility bucket

Designers: F. Eugene Smith Associates
Client: Wooster Brush Co.

Designed for both amateur and professional painters, this bucket, with the addition of a simple screen, becomes a "roller pan." The flat surface enables use on steps or ladder. When handle is in rear position, brush can rest horizontally on it. Finger ledge at bottom aids pouring from non-drip spout. Made of paint-shedding polyethelene, the bucket has six-quart capacity.

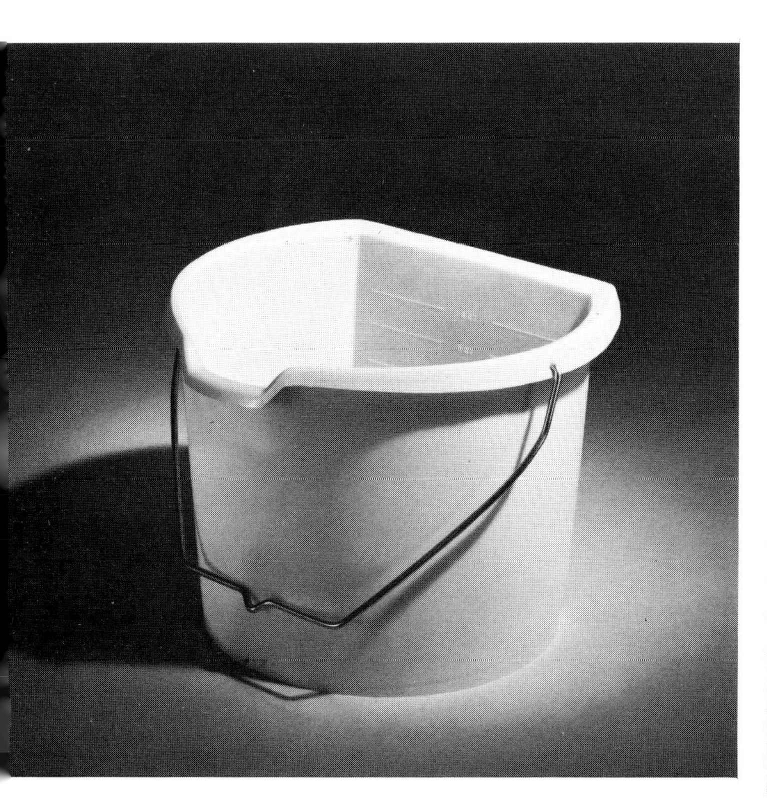

Portable power tools

Designers: Pulos Design Associates, Inc.
Client: Rockwell Manufacturing Co.

Non-conductive housings, with double-insulated construction, eliminate electrical shock. Basic color throughout line is sage green.

"Plane 'R-File"

Designer: L. Garth Huxtable, in collaboration with client's engineering staff
Client: Millers Falls Co.

Versatile do-it-yourself abrading tool files with straight handle, planes when handle is angled. Frame is die-cast aluminum, handle of red phenolic.

Shock-proof impact wrench

Designer: L. Garth Huxtable, in collaboration with client's R & D engineering staff, L. C. Pratt, Director
Client: Miller Falls Co.

One of an extensive line of heavy-duty double-insulated tools with common frame. Die-cast aluminum frame with insulated liner is squared at front for four-bolt connection to gear case, rounded at rear to receive fixed or rotating Lexan back caps. Handles are Lexan.

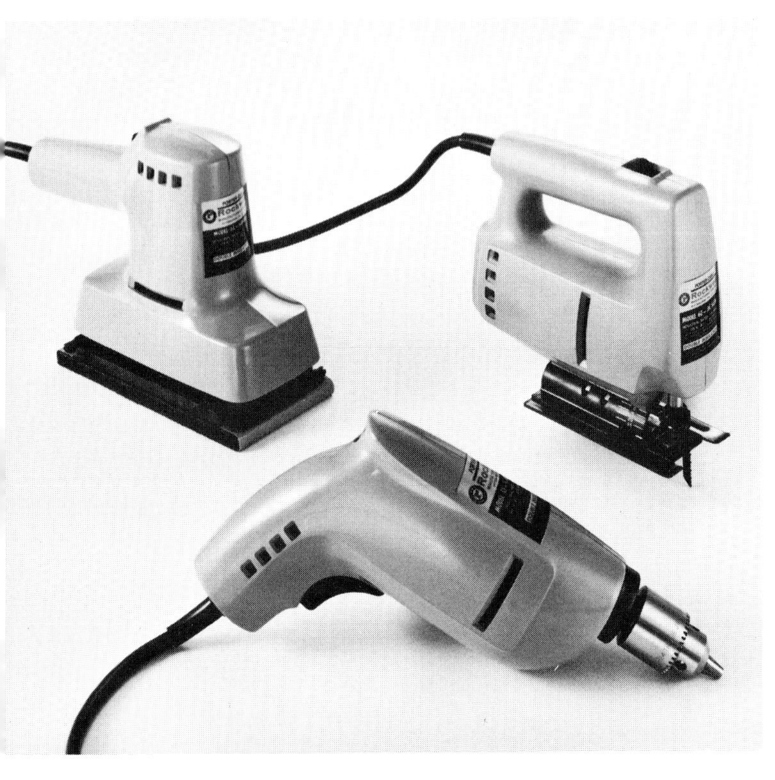

Pop-rivet tool

Designer: William E. Sydlowski
Client: Celus Fastener Corp.

Arms are steel stampings; hand grips are injection molded soft vinyl. Interchangeable nose pieces stored under top grip.

Socket wrenches

Designers: Greenlee-Hess Industrial Design: Hugh T. Greenlee, Roy P. Hess
Client: Wright Tool & Forge Co.

Direction of turn-set bar redesigned for easier fingertip control. Molded rubber grip on larger tools to improve grip and discourage use as hammer.

Hand-held foam shaper

Designers: George A. Beck Associates: Roy Bradbury
Client: Foamold, Inc.

Designed to encourage new uses for expanded bead Styrofoam manufactured by client. Nichrome wire stretched between leading edges is heated by a transformer plug. Spring-loaded switch is in handle. Described as "safe for children over 5."

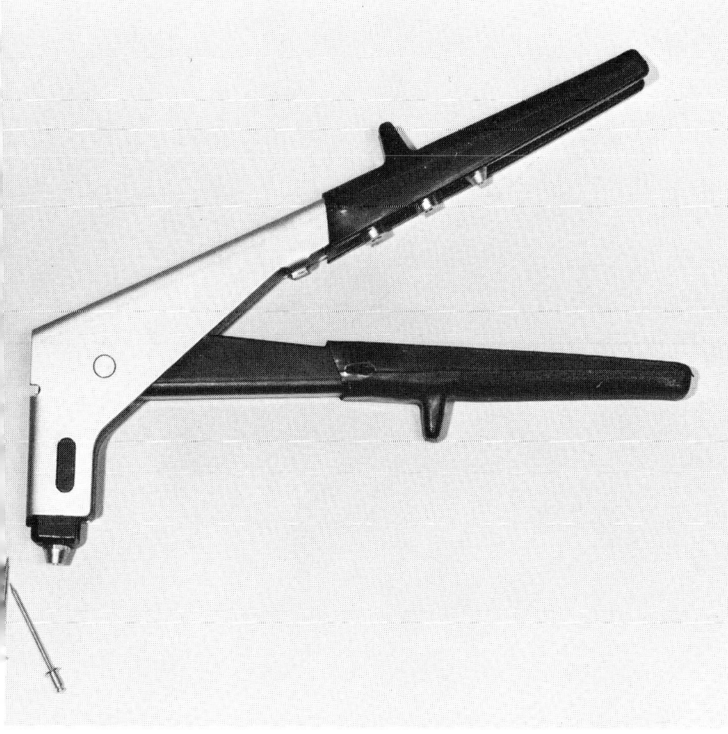

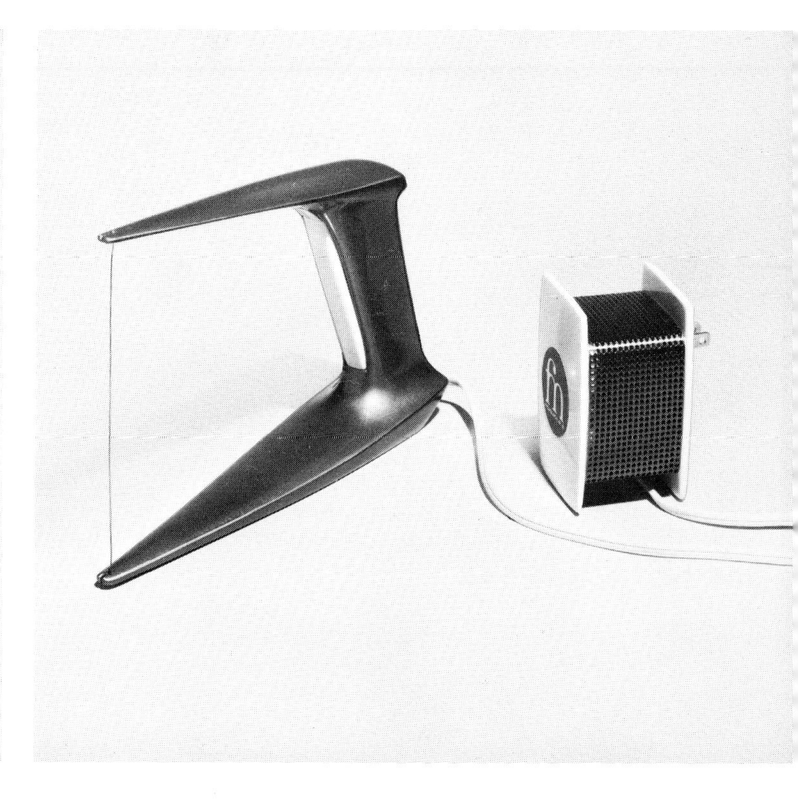

Super-8 movie camera

Designers: Latham Tyler Jensen, Inc.: Dale R. Caldwell

Client: Argus, Inc.

Camera is designed around "flip grip" handle, which provides stable grip for shooting, firm grip for carrying and for pulling the camera from its holster. Flipped up, it locks against camera body. When flipped down, it turns on light meter.

Showmaster super-8 movie camera

Designers: Latham Tyler Jensen, Inc.: Dale R. Caldwell
Client: Argus, Inc.

Emphasis is on ease of use.

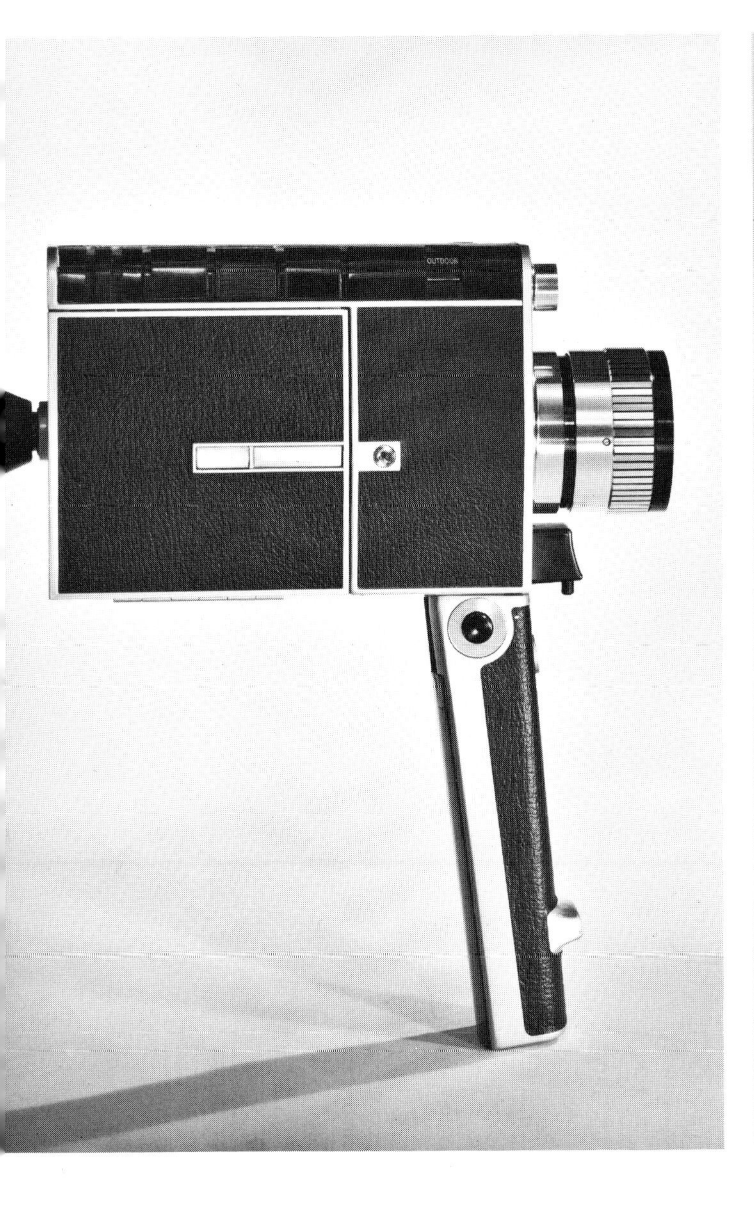

Portable fluorescent light

Designers: Don Doman Associates, Inc.: Don Doman, Tom Hillebrand
Client: Burgess Battery Co.

Battery-powered fluorescent light has 12" tube; can also be run from 110 outlet. Handle and case molded of high-impact polystyrene, dark gray or olive. Stainless steel reflector and trim.

Electric fishing reel

Designers: Harper Landell & Associates
Client: Woodstream Corp.

Main body of unbreakable Lexan with salt-water-proof joints. Removable, rechargeable powerpack contains two nickel-cadmium cells which supply power for full day's fishing.

Hydro Drive marine outdrive unit

Designers: Eliot Noyes and Associates: Eliot Noyes, Ernest Bevilacqua, Robert Graf
Client: Hydro Drive Corp.

Form based on both hydrodynamic requirements and human factors.

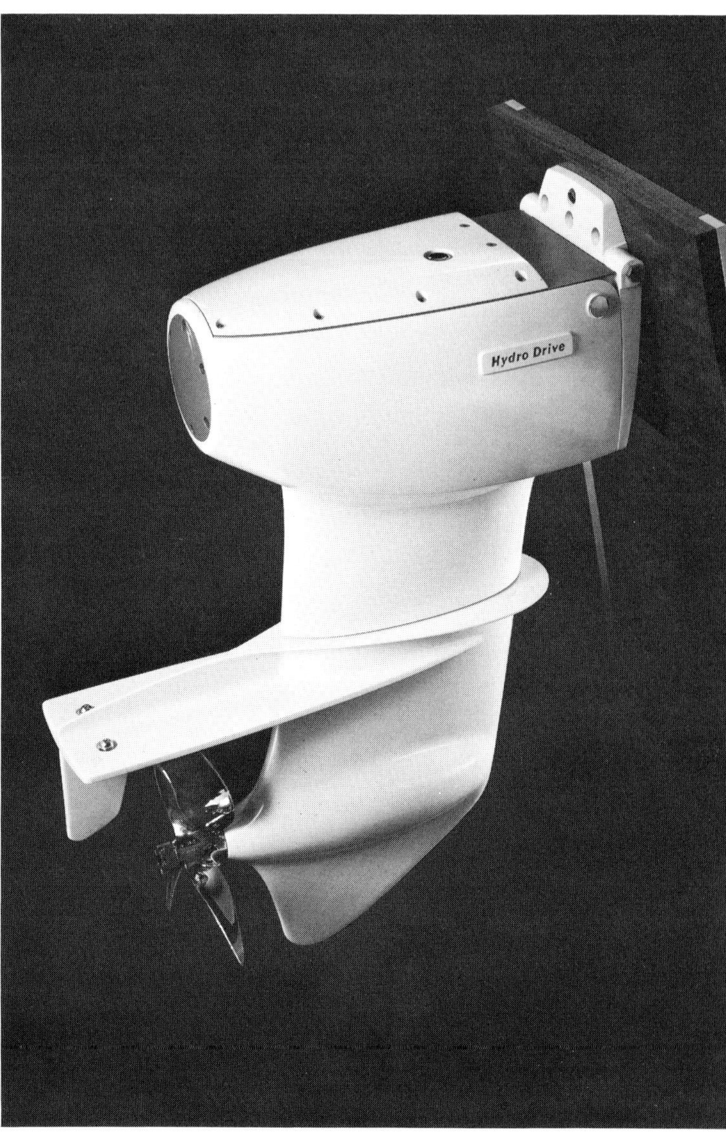

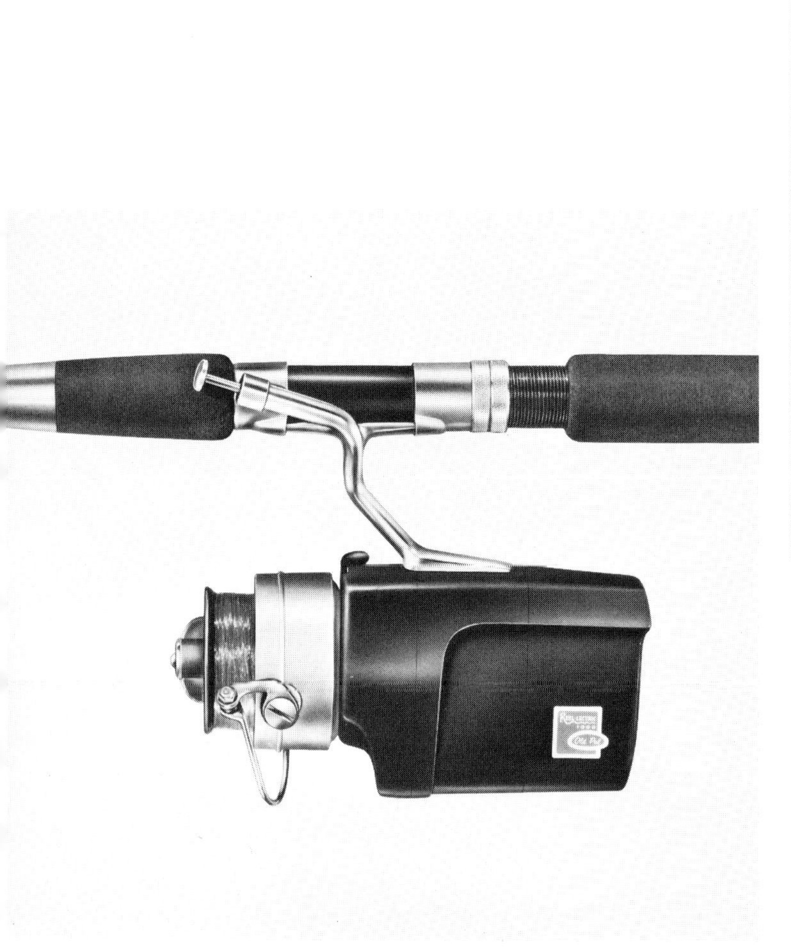

Luggage

Designers: Jon W. Hauser, Inc.
Clients: Hawley Products Co. and American Luggage Works, Inc.

Introduced in 1953, this vinyl-covered, fiberglass reinforced, molded plastic luggage has remained unchanged in design, except for handles and locks, and except for changes of color and pattern as new vinyl treatments have become available.

Luggage

Designers: Design West Inc.
Client: Samsonite Corp.

Extruded magnesium frame used for lightness and strength, also to maintain identity with other of client's products. Materials: vinyl exterior, poly-propylene support, fabric linings.

Attache case

Designers: Design West Inc.
Client: Samsonite Corp.

One of a line of attache cases sold in budget and discount stores; comparable to, but not identical to, the manufacturer's proprietary attache cases. Latches and handle recessed for protection and neatness.

Attache case

Designers: Design West Inc.
Client: Samsonite Corp.

Cost and weight of interior lining are eliminated by injection-molded polypropylene shells with finished interior. Extruded magnesium frame, formed into hoop, provides durable, lightweight structure.

Portable typewriter

Designers: Stevens-Chase design associates: David O. Chase, Philip H. Stevens, Roger Schindler
Client: SCM Corp.

Aluminum and plastics used to reduce weight, subdue noise, effect appearance changes. Top cover slides forward to expose ribbon carrier and spools. Soft white keyboard area eliminates sharp contrast between paper and keys.

Portable typewriter

Designers: Lippincott & Margulies, Inc.: R. G. Smith
Client: Royal Typewriter Corp.

Made to retail for less than $50, this typewriter is aimed at the student market.

Ballpoint pens

Designers: Zierhut, Vedder, Shimano Industrial Design
Client: Paper-Mate Co.

Three pens vary in diameter (.325, .375, .410), permitting choice of grip and width.

Injector razor

Designers: Henry Dreyfuss Associates
Client: ASR Products Co., division of Philip Morris, Inc.

One-piece stainless steel, with serrated plastic underside for grip.

Flashlight

Designers: Stevens-Chase design associates: David O. Chase, Martin V. Maloney
Client: Union Carbide Corp.

Flashlights have generally been cylindrical, or at least symmetrical in cross section, in order to accommodate deep drawing operations. This one, shaped to fit the hand, blends two different elliptical forms with the circular lens end and base. Locking switch button guards against draining batteries. Reverse draw on the end contains all necessary nomenclature.

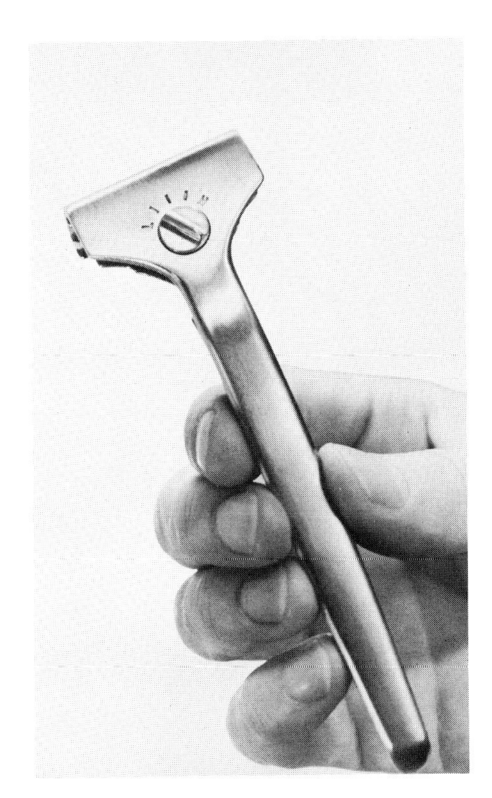

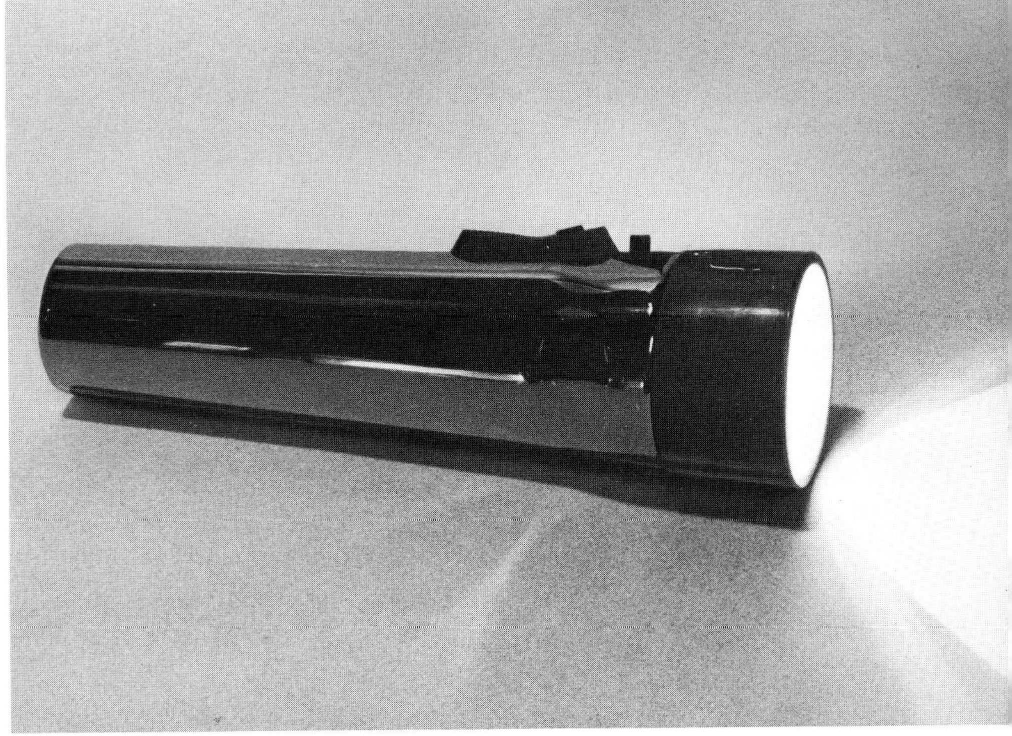

"Avanti"

Designers: Raymond Loewy/William Snaith, Inc.: Raymond Loewy
Client: Studebaker Corp.

Introduced GT performance characteristics and design into the American market, while accommodating American standards of comfort (generous passenger space) and convenience (ample trunk room).

Barracuda Sports Fastback

Designers: Chrysler Corp. Staff: R. G. Macadam, Chief Stylist, Plymouth Exterior Styling; C. G. Neale, Chief Stylist, Interiors and International

Aim was to combine utility with youthful appeal to create station-wagon-like versatility. Rear seat folds forward, opening trunk cavity to the interior.

Dodge Charger

Designers: Chrysler Corp. Staff: C. G. Neale, Chief Stylist, Interiors and International

Simple basic shapes and trim; major curve movement at rear gives sense of power to form.

Mercury Cougar

Designers: Ford Motor Co. Design Center: Eugene Bordinat, VP, Product Planning and Design; A. B. Grisinger; L. D. Ash; J. L. Darden; H. N. Koto; J. P. Aiken; J. F. Van Tilburg; G. L. Halderman; H. M. Tod; D. F. Kopka; D. C. Woods
Client: Lincoln-Mercury Div., Ford Motor Co.

Full-accessoried version of Ford's intermediate sports car, distinguished by fine-edged sheet metal peak running forward from "C" pillar.

Mustang (bottom)

Designers: Ford Motor Co. Design Center: Eugene Bordinat, VP, Product Planning and Design; Joseph Oros; L. D. Ash; D. C. Woods; John Najjar; J. B. Foster; R. H. Wieland
Client: Ford Div. of Ford Motor Co.

Short rear deck and elongated hood contribute to design of low-cost car with sports car appeal to younger market. Flat, low form with hard-edge emphasis adds elegance, some formality.

Lincoln Continental four-door sedan

Designers: Ford Motor Company Design Center—Eugene Bordinat, Product Planning and Design; A. B. Grisinger; D. R. DeLaRossa; R. H. Maguire; L. D. Ash; A. J. Middlestead; R. N. Conrad.
Client: Lincoln-Mercury Div. of Ford Motor Co.

Body-side sheet metal effectively unadorned except for name in script. Chrome molding traces beltline shape. Dual headlights in individual bezels are recessed in convex checkerboard-pattern grille.

Corvair Monza

Designers: General Motors Corp. Styling: William L. Mitchell

Compact concept based on rear engine and unit-construction body. Shell featured thin pillars and relatively large glass areas, uncluttered surfaces.

Corvette

Designers: General Motors Corp. Styling: William L. Mitchell

Second version of mass-produced fiberglass-bodied sports car to be manufactured in this country. Doors open into roof for easy access.

Buick Riviera

Designers: General Motors Styling Staff: William L. Mitchell

High side glass and "turned-in" rocker are used to achieve sharply sculptured lines on upper body, slim appearance on lower. Vertical "nacelles" in forward surfaces of fenders contrast with horizontal grille aperture.

The Consumer in Business and Industry

You always find good design in work things.
—Saul Steinberg

When there is a job to be done efficiently, in a situation in which time and expense are synonymous, the pressures of fashion and status give way to pressures that make for more sensible design. Consequently, design for business and industry tends to be less arbitrary than design for the home. Such considerations as safety, speed of operation, ease of maintenance and reliability almost always take precedence over aesthetic considerations in such cases. The function of a tool, whether the tool is a microphone or a semi-conductor oven, is to enable work to be done. Anything else is secondary.

This, however, is not to say that appearance and other aspects of sales appeal are, or can be, ignored in the design of business or professional experiment. As it turns out, the solution to practical problems frequently has "aesthetic payoffs," as evidenced by the marine diesel engine (page 73), the Shure microphone (page 89), the "Dispens-O-Disc" system (page 94).

But if a design truly integrates form and function, it is difficult ever to isolate aesthetic factors. Most of the equipment for electronic data processing (pages 78 to 88) was designed with certain objectives in common. A vastly complex system had to be simplified and its operation clarified. Impersonal machinery had to be made to look humanly manageable, yet business-like. Calculating equipment had to be revealed as what in fact it is: an extension, however remote and abstruse, of human calculations.

For all that (often, *because* of all that), how handsome some of it is! Nor is its handsomeness irrelevant to the client's basic interest. Since the machinery itself figures prominently in display windows, in expositions, and in the many visible applications of computers, it, like consumer products, sells in part because of how it looks. That is not entirely surprising, since the appearance of business equipment can affect the morale of workers and customers alike.

But isn't the sales appeal of capital equipment more solidly based than the sales appeal of goods in heavily competitive consumer areas? It may be true, as package designers were fond of declaring a decade ago, that women walk through supermarkets in a physiologically measurable state of trance. Yet, surely if a purchasing agent goes through life that way, he and his company are in trouble.

Yes. But behind the shopper's trance is the fact that the supermarket confronts her with dozens of sets of choices, each of them requiring that she choose between products claiming to be "equally superior." Without a mobile laboratory, or pocket index to *Consumer Reports,* how is she to tell which really is superior? Or if, as is very often the case, all of them are the same, what is her basis for choosing?

Just as the designer provides a basis for the housewife's choice, he also provides one for the purchasing agent or corporate executive. A designer of military equipment observes that the officers who buy this equipment select items largely because of the way they look. It is not a frivolous reason, and their situation is not unlike the housewife's. The men who make buying or leasing decisions about computers are not likely to know any more about electronic circuitry than the women who buy detergents know about enzyme destruction. Once they are assured that the product can do what it is supposed to do (an assurance that comes from salesmen, from advertising, from the maker's reputation, from colleagues), they buy on faith—faith buttressed by appearance.

Other supposedly amateur values find their way into commercial equipment. The first electric carving knives were awkward to use because the motor and the housing were too large for the hand to grasp comfortably. By moving the motor off the direct axis of the blades, lowering it and angling it, the designers of the Hamilton Beach knife made the handle an easily gripped extension of the blade. The object was to design a product for at-home carving (page 24). The product was extremely successful in the home market, and was followed by a commercial version for professional cooks and countermen.

While business and industrial equipment is ordinarily not designed for elegance, there are cases in which elegance is exactly its function. A luxury restaurant is a perfect example, and there are few more luxurious than the Four Seasons in New York. The restaurant itself was designed to look as though it belonged in the Seagram Building, which happens to be where it is. The services, of which the silver service pots on page 15 are typical, were designed to look as though they belong in the restaurant.

Although the commuter car on page 74 is not produced in mass, it is produced for masses of people. And its interior, like other public interiors, is designed to anticipate human behavior and influence it. While the seating arrangement tries to make at least the illusion of privacy possible, it also acknowledges the passenger's tendency to take the inside and outside seats. The aisle seat in this car has a shortened back to accommodate a hand rail for standees and/or walkers through, but also to "merchandise" the unpopular center seat.

In the design of work equipment, certain criteria have long been paramount. The straddle stacker on page 75 has a handle designed to reduce operator fatigue, prevent injuries, save time, and help minimize load damage. It is no longer unusual for a concern with overall efficiency to be expressed in plant equipment. The *office,* however, has rarely been exposed to this kind of treatment. Except for the stenographic posture chair, and some debate about whether desk drawers help or hinder executives, office furniture has been regarded pretty much as just a matter of taste. But a few years ago the director of the Herman Miller Company's research division, Robert Probst, began asking questions about how people work in offices. Sitting or standing—or even lying? And for how long? Should a door be opened or closed? Should all offices even *have* doors? Where should phones be located? How often do executives nap in offices? Should the practice be discouraged?

The point is not that these questions had never been asked, or in some measure answered, before. But they had never been thought answerable by the design of furniture. The answer in this case, designed by George Nelson & Co., is the "action office" (page 104)—a modular office system adaptable to any number of working configurations, and providing work surfaces, phone stations, quickly accessible storage, and the display space for work in progress.

Fire hydrant

Designers: Cornelius Sampson & Associates: Donald F. Baldocchi, Cornelius Sampson
Client: East Bay Municipal Utility District

Designed to harmonize with contemporary architecture, this hydrant costs no more to produce than conventional hydrants, and is planned around existing pipes and valves, with fittings that accommodate standard fire department wrenches and spanners.

Caster

Designers: Don Dailey Associates: Don Dailey
Client: Faultless Caster Co.

Medium-duty caster for teacarts, television stands, and other light furniture. Unsightly "yoke-horn" is eliminated by use of die-cast internal "horn" sandwiched between plastic discs.

Water fire extinguisher (hand portable)

Designers: Latham Tyler Jensen, Inc.: Frederick S. Brennan

Fiberglass receptacle with molded rubber end caps is departure from conventional stainless steel, brass and bronze fabrication. All handle parts are injection-molded plastic. Advantages: no corrosion, simpler manufacturing techniques, opportunity for architects and builders to specify colors and finishes.

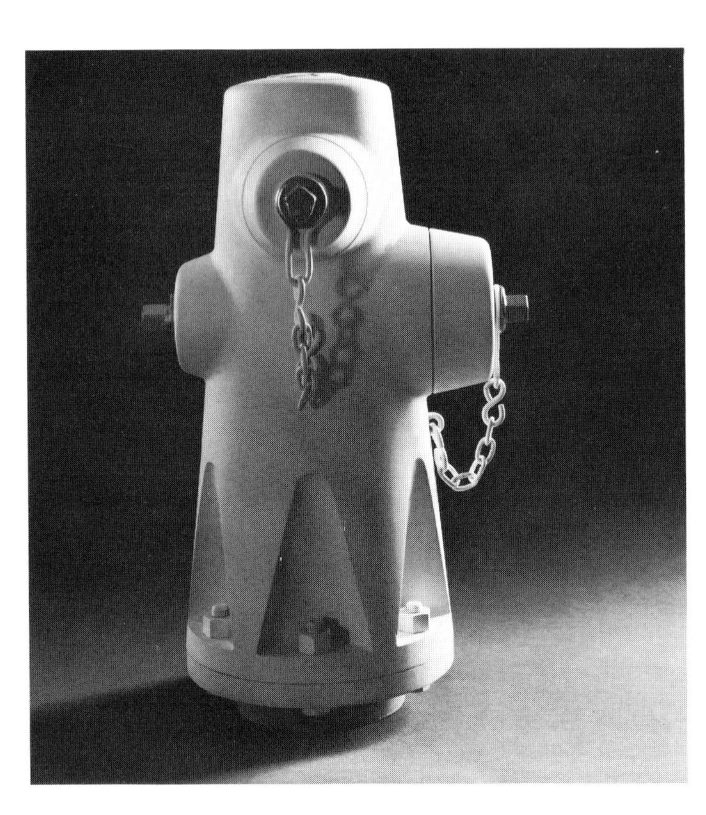

Vault door

Designers: Henry Dreyfuss Associates
Client: The Mosler Safe Co.

Laminated metals, with square corners, achieve manufacturing economies without loss of security.

Gasoline pump

Designers: Eliot Noyes and Associates: Eliot Noyes, Ernest Bevilacqua, Robert Graf
Client: Mobil Oil Corp.

New form results from reexamining functions of pump and duties of attendant.

Barber's clippers

Designer: Alfred W. Madl
Client: John Oster Manufacturing Co.

Barber's clippers tend to be awkward to use and fatiguing; and they are noisy, vibrate a lot, and become excessively hot. The aim of this design was to produce an efficient clipper without these faults. Various different plastics were used, including polycarbonates for the housing.

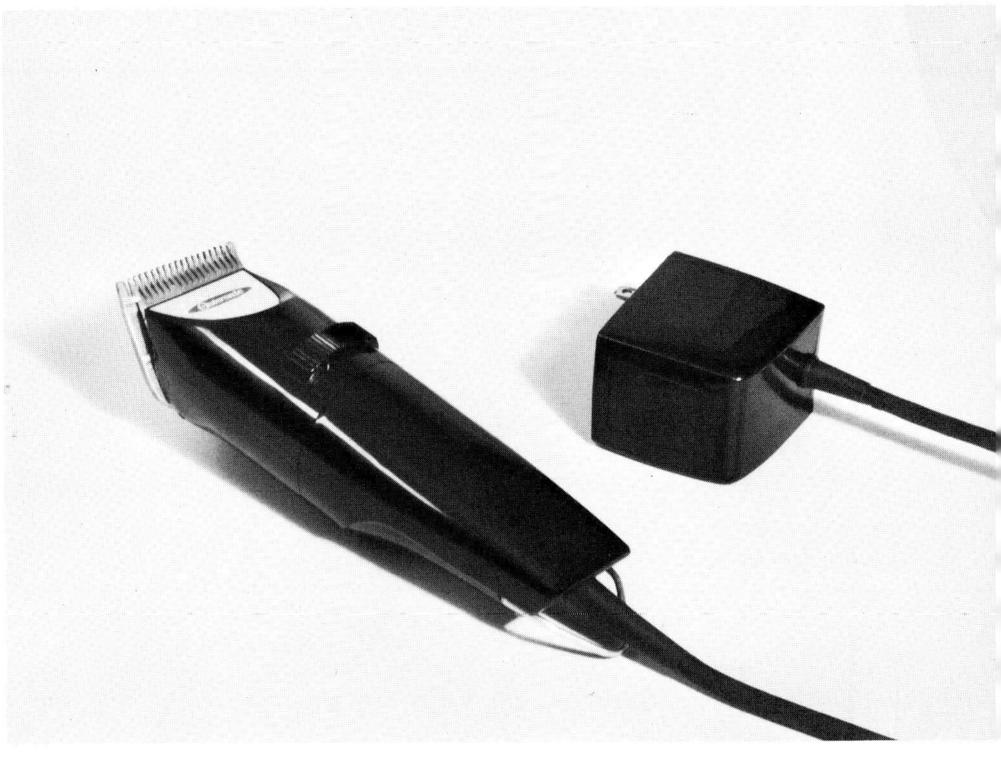

Web offset printing press

Designers: Donald Deskey Associates
Client: American Typefounders Co.

Aluminum guards are adaptable to two other press sizes.

High-voltage light generator unit

Designers: Zierhut, Vedder, Shimano Industrial Design
Client: Marshall Laboratories

Console on casters addresses problem of heavy equipment requiring some mobility. Cabinet is mostly sheet-metal construction with square steel tube framework. Upper part is sand cast, with sheet-metal control panel.

Semi-conductor oven

Designers: Dana Mox Associates
Client: Lindberg Hevi-Duty, division of Sola Basic Industries

Functional parts are easily reached from front for servicing. Fabrication is modular, adaptable to various sizes.

Diffusion furnace

Designers: Dana Mox Associates
Client: Lindberg Hevi-Duty, Div. Sola Basic Industries

Modular design permits one-stack unit to be expanded into two or three stack model as desired. Solid-state controls have slide in-slide out accessibility.

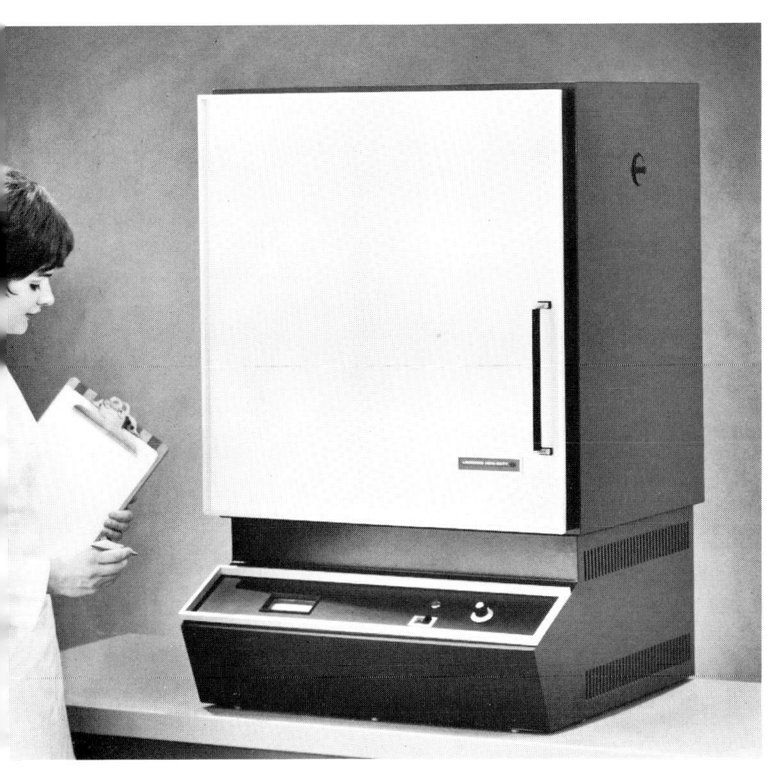

Servofeed automated turret lathe

Designers: Henry Dreyfuss Associates, Inc.
Client: The Warner & Swasey Co.

Control panel slides along rail, following operator as he makes adjustments.

Electric gearshift drive

Designers: Walter Dorwin Teague Associates, Incorporated
Client: Lima Electric Motor Co. (subsidiary of Condec Corp.)

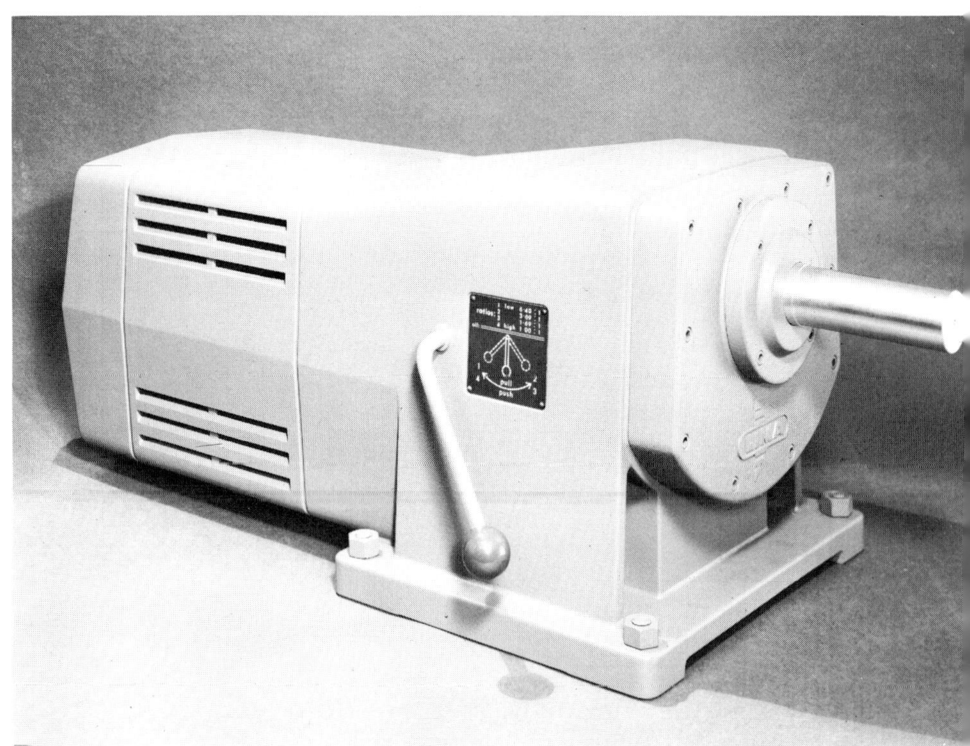

Electronic siren-light

Designers: Stevens-Chase design associates: Philip H. Stevens, David O. Chase, Alan W. Brownlie
Client: R. E. Dietz Co.

Combination is less expensive than separately purchased light and siren, and installed in half the time. Protrusion at rear houses horn drive, providing for simplified tooling and molding. Unit made of molded reinforced fiberglass plastic.

Hydraulic adjustable speed drive

Designers: Wilson, Hopkins, Fetty & Kitts: Harold D. Fetty, Richard E. Watson
Client: Vickers, Inc.

Hydraulic components are combined into simple configuration, integrated with standard electric motor. Highly viscous stippled blue-grey paint unites die-cast fan shroud with other cast iron components.

Carbon deleaver

Designers: Dana Mox Associates
Client: Uarco, Inc.

Low-cost, limited-space machine affords unskilled operator a quick visual check of forms being deleaved.

Automatic burster-stacker

Designers: Dana Mox Associates
Client: Uarco, Inc.

Operates in-line with data processing printers, adjusts automatically to printer's speed.

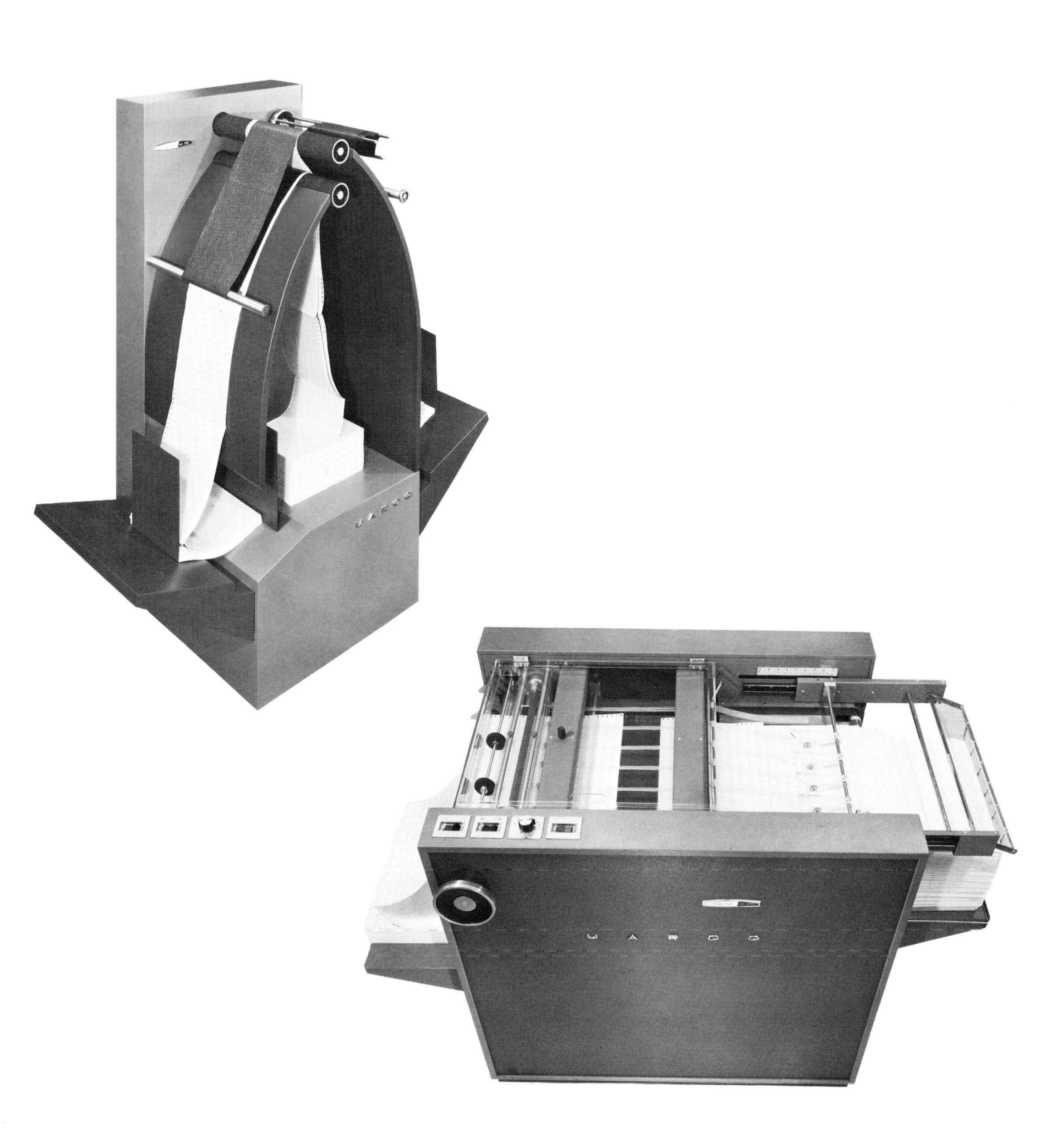

High-speed carbon deleaver
Designers: Dana Mox Associates
Client: Uarco, Inc.
Removes up to three carbons from continuous-control punched forms.

Forms-handling equipment
Designers: Dana Mox Associates
Client: Uarco, Inc.
Family line of similar devices, all compatible in design character and using identical roll form section.

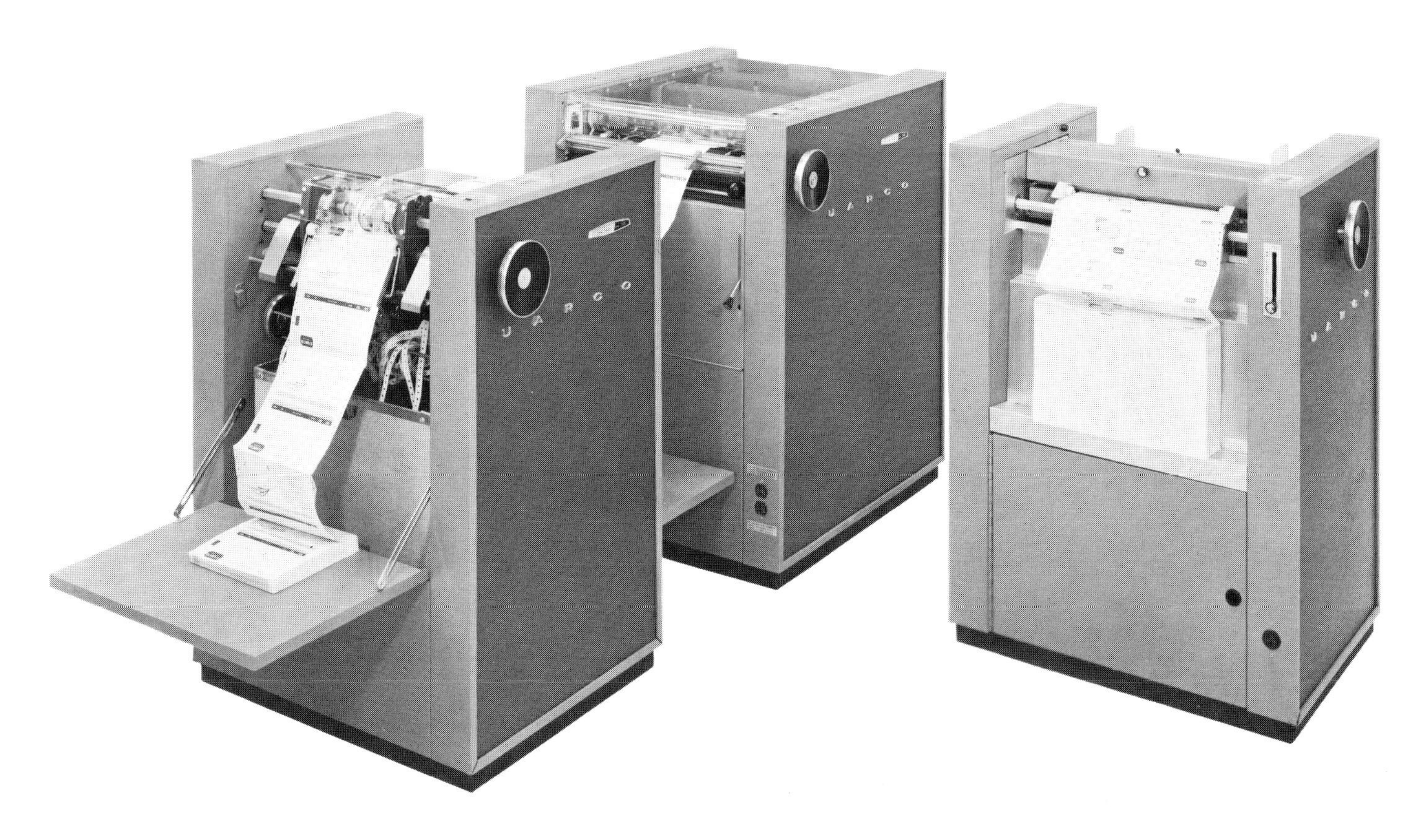

Portable suction dredge

Designers: Arnold Wolf Associates: Darrel A. *Lauer,* in collaboration with client's staff
Client: PACECO

The designers describe the project this way: *"Problem: The Bihar River in India overflows during monsoon seasons, changing course and flooding the surrounding farmlands. The Bihar government required a small portable dredge to dig a permanent channel for effective flood control. 'Portable' in this case meant that the dredge would have to be disassembled into pieces small enough for transport on narrow-gauge railroads. The unit was to be self-propelled instead of relying on tug-boats, and the remote areas of operation and extended periods of work required accommodations for a crew of eight. Solution: The portability requirement dictated an unusually narrow hull section. This limitation, combined with the extra propulsion equipment, demanded extreme care in the planning of the machinery space and adjoining crew's quarters (which include an all-electric galley). Equally precise human factors analysis was applied to the cab design, which affords the operator 360° visibility and centralized control."*

Marine power center

Designers: Van Dyck Associates
Client: Harvey Hubbell, Inc.

Form derives from necessity of keeping rain water from the electrical outlets. Door detail expresses full weather gasketing, permits full access to interior while unit is being installed. Unit accepts existing Hubell spring-loaded outlet cover plates.

Marine diesel engine

Designers: Eliot Noyes and Associates: Eliot Noyes, Ernest Bevilacqua, Robert Graf
Client: Cummins Engine Co., Inc.

Forms have been simplified for casting, accessibility, cleaning, appearance of power.

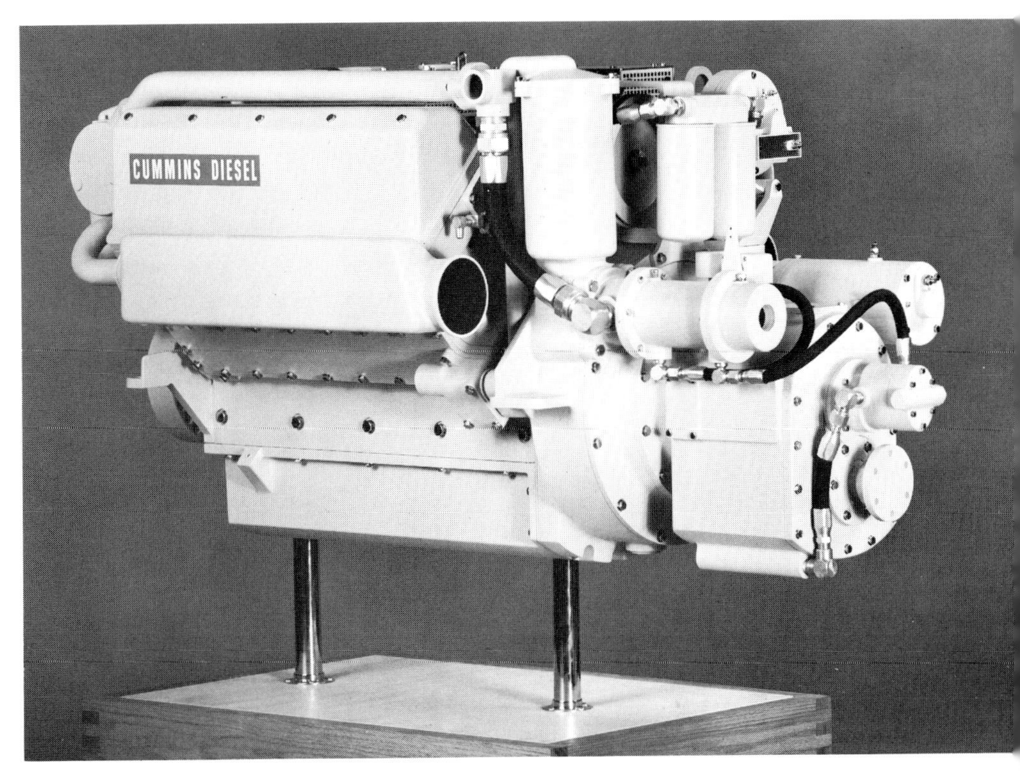

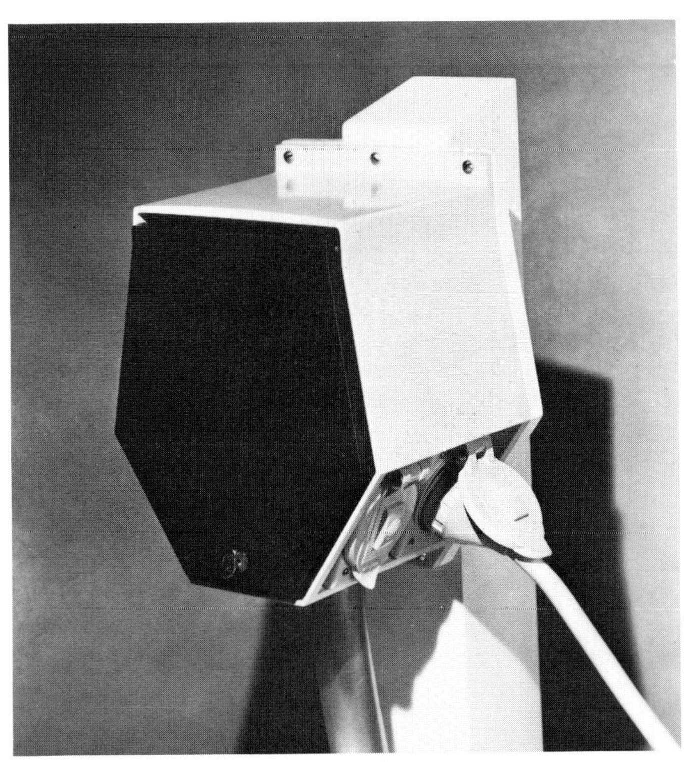

Long Island Railroad commuter car

Designers: Sundberg-Ferar, Inc.
Client: Metropolitan Commuter Transportation Authority

Seats arranged two on one side, three on other, with contouring and individual knee wells to give passenger sense of "owning his own space." Aisle seat on three-seat side has no head rest, for effect of openness; also to encourage use of center seat.

Heavy-duty straddle stacker (bottom)

Designers: Richardson/Smith, Inc.: David D. Tompkins, design coordinator
Client: Crown Controls Corp.

Operator's handle has wrap-around guard to protect user's hands, and contact-actuated "gut-button" brake bar, in case he backs himself against a wall. Entire mechanical and hydraulic systems can be exposed for service by removing two bolts and swinging body shells open.

Pallet truck

Designers: Richardson/Smith, Inc.: Terry J. Simpkins, Jack L. Petrick, David B. Smith, Deane W. Richardson, in collaboration with Harold Stammen, client's project engineer
Client: Crown Controls Corp.

Rearrangement of power drive and lift components result in smaller truck. Light beige replaced standard orange and yellow, while preserving safety-visibility features of those colors.

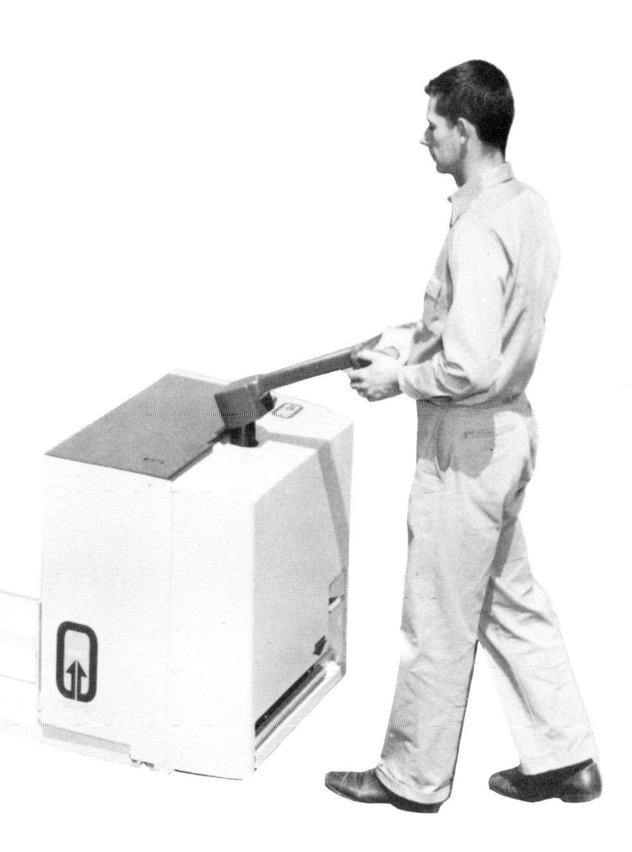

Embankment compacter

Designers: Henry Dreyfuss Associates
Client: Hyster Co.

Two engines drive separate compaction modules.

Towing tractor

Designers: Clark Equipment Co. staff: Kent Brown

Since tight tooling budget did not permit precision sheet metal fit-ups, "mismatches" are incorporated into design. Hood formed around engine increases visibility, rear fender caps are inexpensive iron castings which help counter-weight drive wheels. Push-bumper plate and grille are flame cut.

Lift truck

Designers: Henry Dreyfuss Associates
Client: Hyster Co.

Tilt cylinders are integrated with overhead guard to reduce mechanical effort, provide sturdy, safe protection for driver.

Computer console

Designers: IBM staff: W. S. McCormick, W. F. Kraus, industrial design manager; Eliot Noyes, consultant design director

Since operator activity is only intermittent, all operator-interaction items have been elevated and console is designed at "sit-or-stand" height. Basic materials are stainless and formica-laminated steel.

Computer console

Designers: Honeywell EDP staff: John F. Graham, manager, industrial design

Central electronics cabinet with cantilevered pedestal-mounted keyboards on either side. Channel construction that supports pedestals and connects them to cabinet also encloses cable raceway. This enables all electronics to be housed in cabinet, freeing pedestal design for remote terminals.

Direct-access storage facility

Designers: IBM staff: Donald H. Wood, Donald A. Moore, manager; Eliot Noyes, consultant design director

Disk drawers open to the front for disk-pack changing, and to the rear for servicing. Prior to sliding a module out the rear for servicing, drawer front is automatically transferred from drawer to cabinet frame, thus keeping front appearance unchanged while file is being serviced.

Digital computer

Designers: Digital Equipment Corp. staff: James Jordan

Size reduction made possible by integrated circuits.

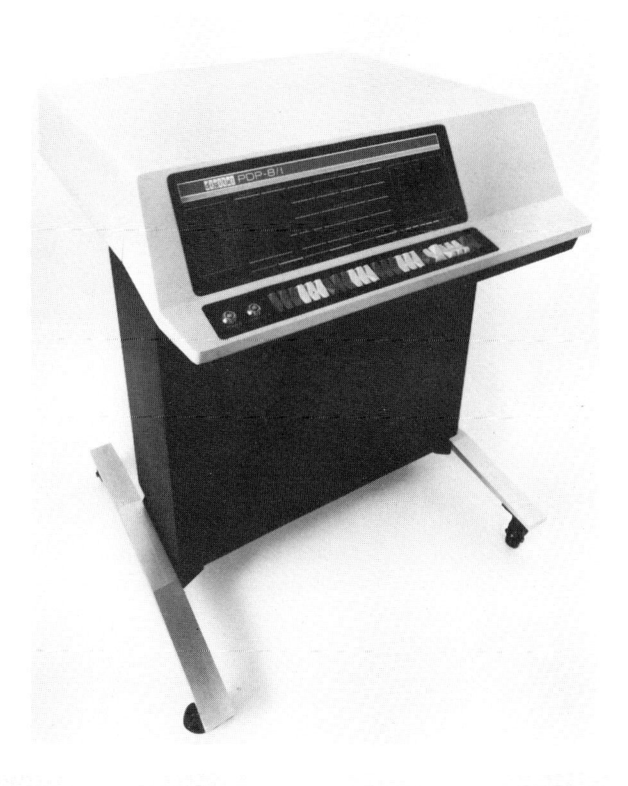

Programmed data processor

Designers: Van Dyck Associates
Client: Digital Equipment Co.

Power supply in bottom unit; racks open book-like to expose wire wrapping for service and modification. Upper cabinet has translucent covers.

Optical reader

Designers: IBM staff: J. K. Hockenberry, E. J. Sabella; Eliot Noyes, consultant design director

Optically scans cash register tapes, transmitting data directly to computer. Covers finished in textured vinyl. Operator panel made of brushed stainless steel, exposed hardware is polished chrome.

Bank terminal

Designers: IBM staff: Edward R. Wiener, Ray B. Wheeler, I. D. manager; Eliot Noyes, consultant design director

Molded acrylic journal tape window permits visual checking and partial correction capability, while preventing access to tape. Extension of same window protects type ball, reveals print line, and provides acoustical baffle.

Optical reader

Designers: IBM staff: Collan B. Kneale, Ray B. Wheeler, I. D. manager; Eliot Noyes, consultant design director PJ-8

Same basic machine as on facing page, except it is larger and reads various sized documents.

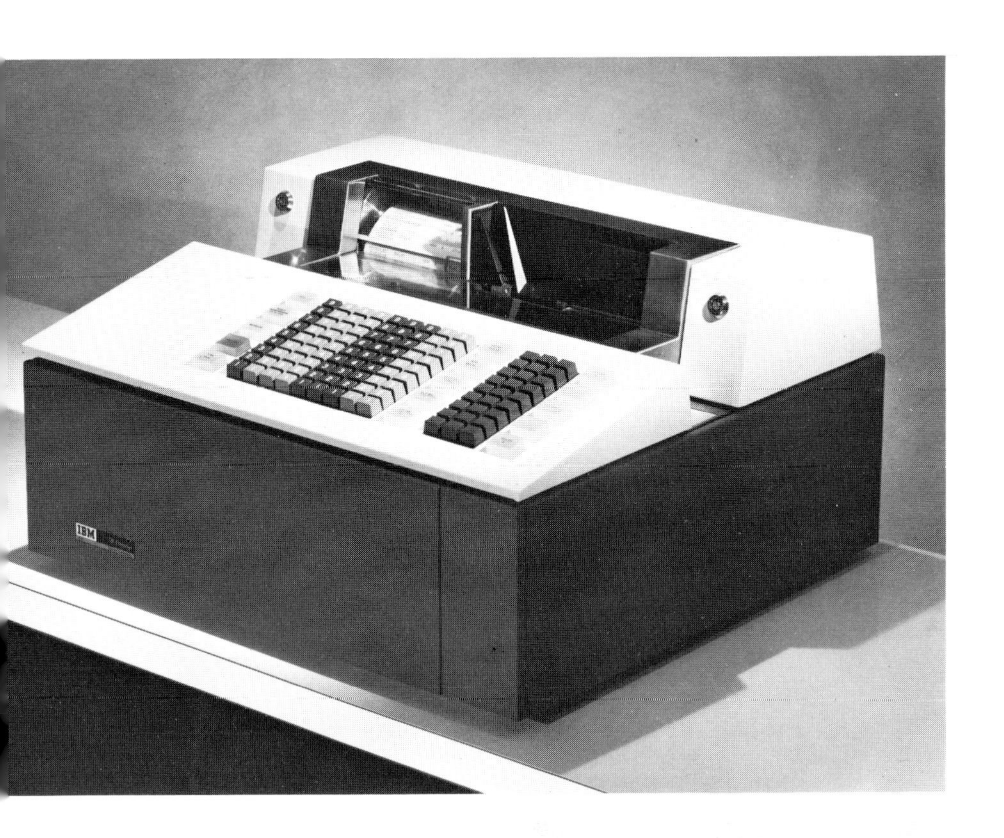

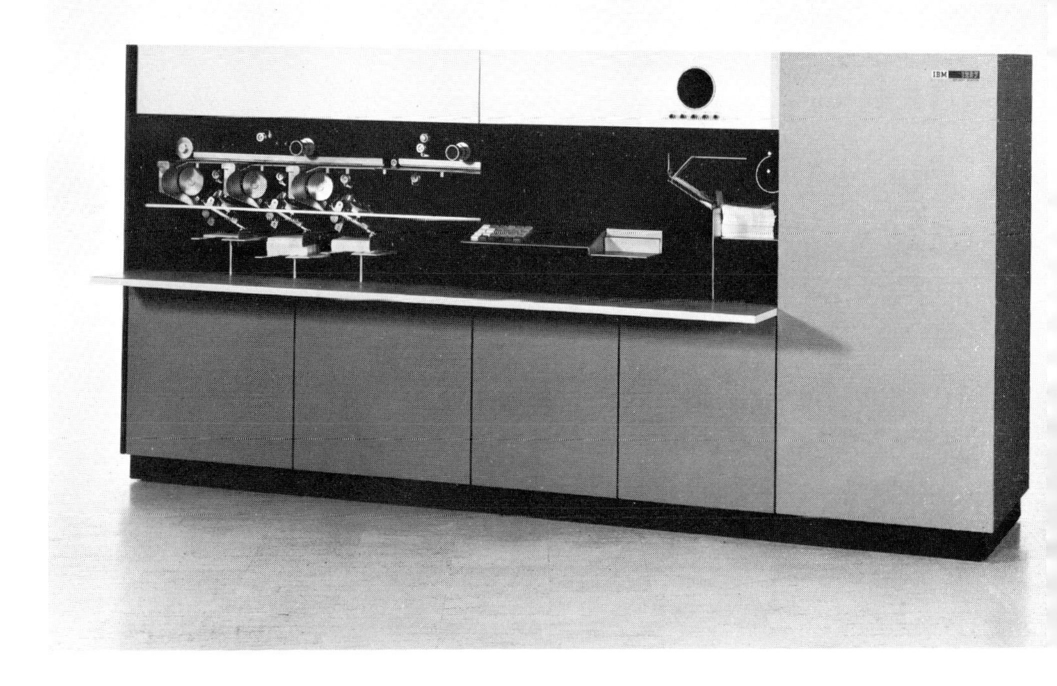

Remote station

Designers: Ampex Corp. staff: Robert W. Bornschlegel, Sr. Industrial Designer; Frank T. Walsh, Director, Corporate Industrial Design Dept.

Control station for slow motion video system has desk-top keyboard, time display at far end for good viewing angle.

Desk-top electronic calculator

Designers: Zierhut, Vedder, Shimano Industrial Design
Client: Friden, Inc.

Except for keyboard, product has no moving parts; physical elements are almost purely electronic. Arrangement of circuitry results in high silhouette, visually lowered by horizontal break in mass.

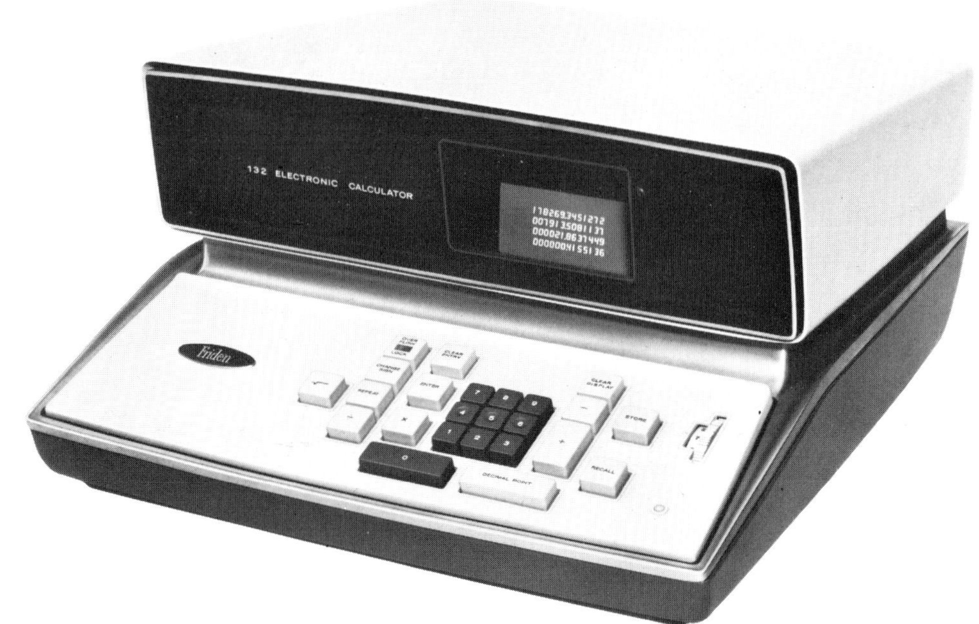

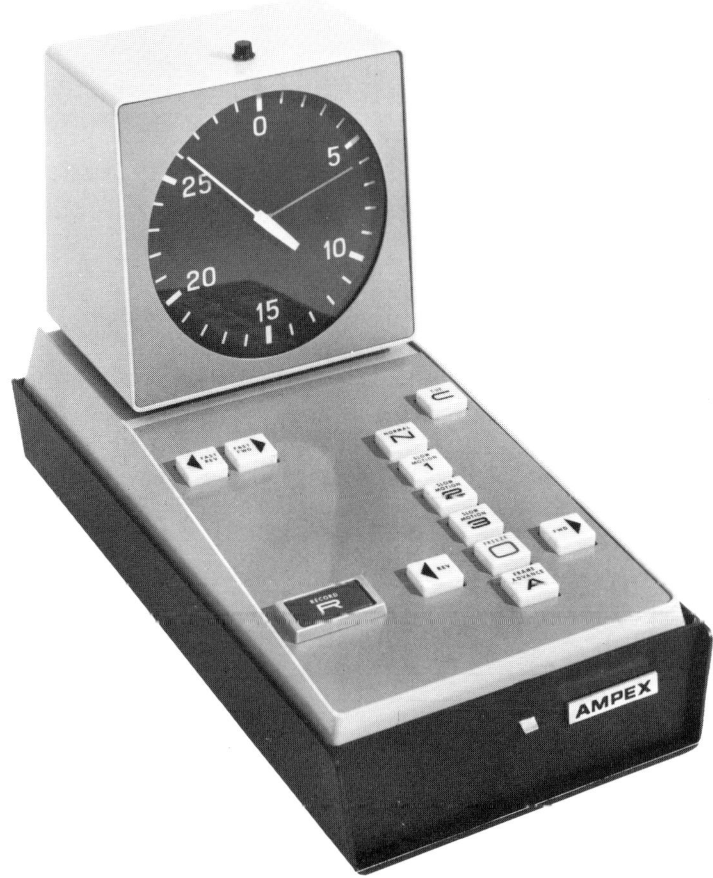

Modular keyboard system

Designers: Honeywell EDP Staff: John Graham, Manager of Industrial Design
Client: Micro-Switch, Div. of Honeywell

Modular low-cost keyboard system for computer input devices and other information-handling uses. Two-piece button approach permits great variation in shape and color, and components may be arranged into any number of matrices adaptable to custom designs, prototypes, or production quantities. In the case of the latter, panels and consoles can be rapidly reconfigured to incorporate changes in design.

Desk-top unit (bottom)

Designers: IBM staff: Frank Wilkey, Jr., Walter Furlani, I.D. manager; Eliot Noyes, consultant design director

Semi-portable, cable-connected desk-top unit. Service requirements met by mounting all components to exposed central frame and applying snap-on plastic covers.

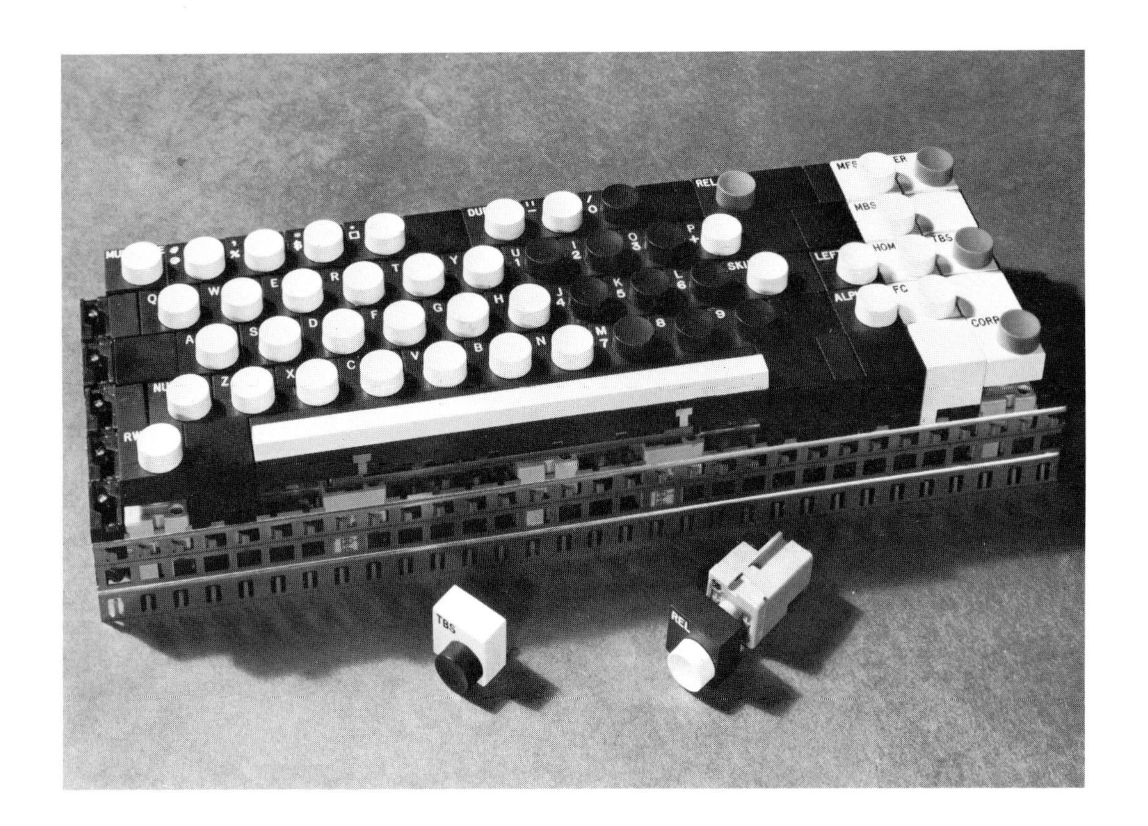

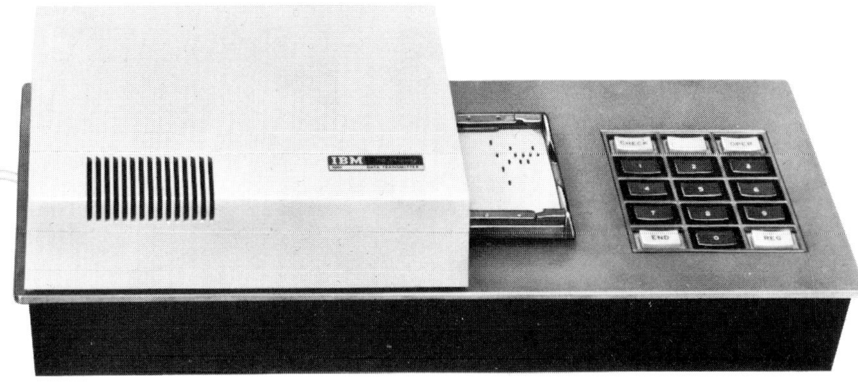

Data station

Designers: Honeywell EDP staff: John Graham, manager, industrial design

Modular 22-inch cube houses various mechanisms, allowing flexibility of arrangement for varying floor spaces, and minimizing number of parts. Keyboard rotates and tilts into position operator finds most comfortable.

Electronic enclosure system

Designers: Hewlett-Packard staff: Carl Clement, design director; Tom Lauhon, Allen Inhelder, Andi Are, Don Pahl, associates

Number of electronic instrument cabinet sizes was reduced from 65 to 16 by this integrated system of interchangeable "erector set" parts. System satisfies varying functional requirements of bench, rack, and field use.

Display station

Designers: IBM staff: S. Cottier, design project manager; W. F. Kraus, I.D. Manager; Eliot Noyes, consultant design director

Keyboard is removable from main unit, which can be used separately as display device only.

Microfilm reader-printer

Designers: Harley Earl Associates: S. M. Highberger

Client: Microfilm Products Div., 3M Co.

Brushed aluminum trim in hard-wear areas is combined with finely textured vinyls to suggest high-calibre photographic capability. Recessed control knobs facilitate hand movement along front control panel. Front castings pull out to make complete film track accessible. Instrument is cooled by air flow through base side screens. Combined hood side and top panel hinges up for replenishing developer fluid.

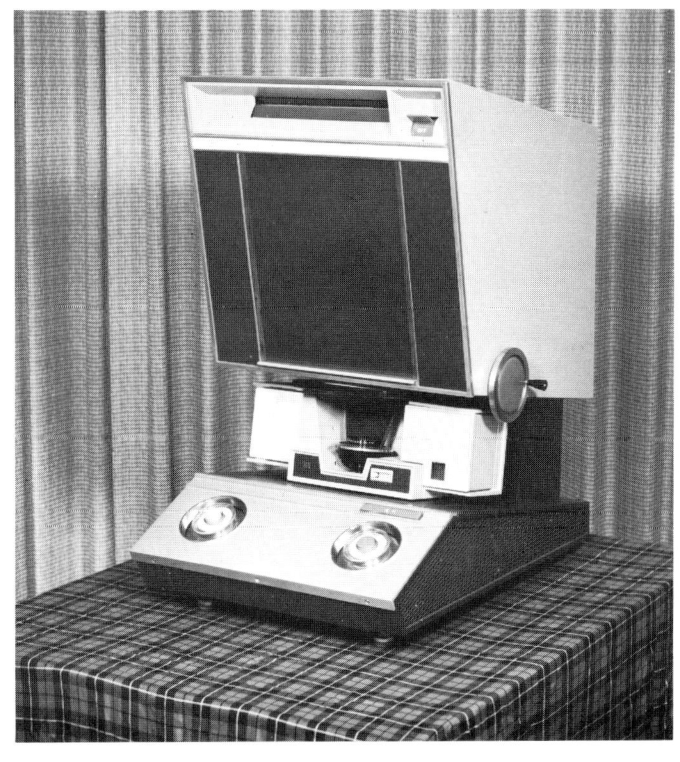

Video tape recorder system

Designers: Deschamps-Mills Associates: Richard S. Hart (recorder), Raymond Massaccesi (cart).
Client: Revere-Mincom, Div. of 3M Co.

Made for industrial and school use, the recorder requires rugged enclosure: its die-cast deck and frame are mounted in a vinyl-covered wood box. Cart makes it a compact, mobile unit with a-v monitor, camera, and storage area for cords, tripods, tape, accessories.

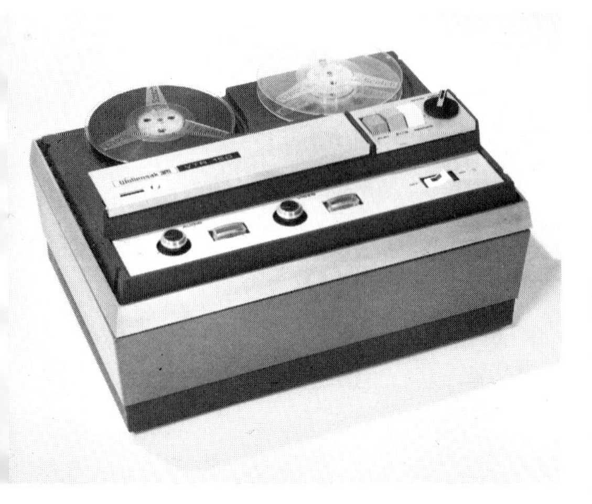

Vertical process camera

Designers: Herbst-LaZar Industrial Design: W. Herbst, R. LaZar, A. Nagele
Client: W. A. Brown Manufacturing (division of APECO)

Cantilevered copy-board housing moves on telescoping track, reducing camera's length by one-third.

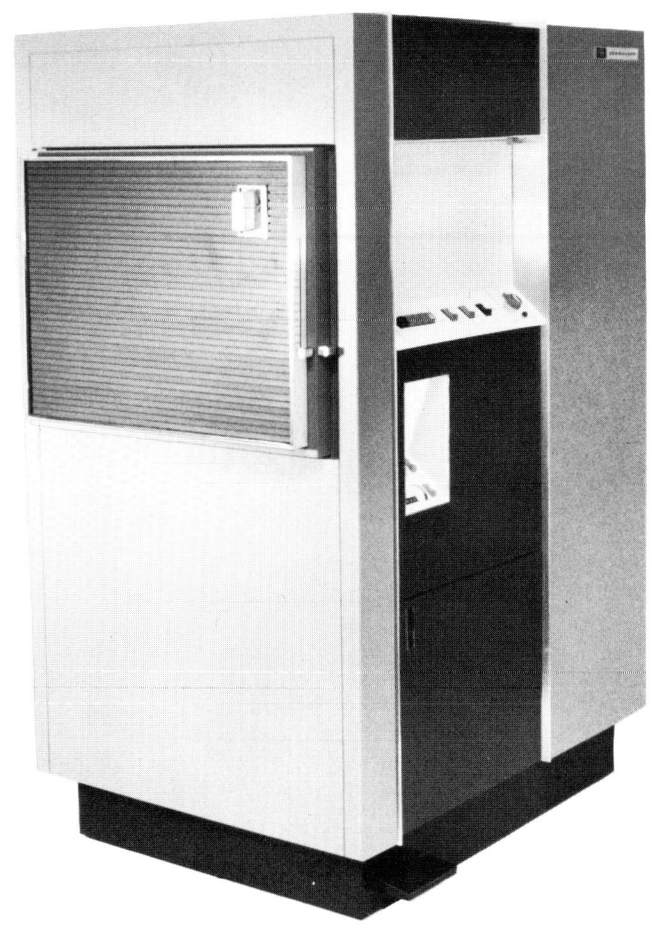

Color video recorder

Designers: Ampex Corp. staff: Gene Bozarth, Sr. Industrial Designer, Don L. Johnson; Frank T. Walsh, Director, Corporate Industrial Design Dept.

Compact broadcast-quality recorder can be mounted in back of small truck or microbus. Vinyl-clad surfaces reduce scuffing.

Mobile/portable data recorder

Designers: Tepper + Steinhilber Associates, Inc.: John Hayashi, Gene Tepper
Client: Pacific Electro Magnetics Co. (PEMCO)

Instrument must operate at any angle (upside down when airborne) and under severe conditions. Weighs 60 lbs., has cast aluminum end bells, is adaptable to 19" rack mounting. All major sub-assemblies are plug-in units. Window permits visual check of capstans, with cover on.

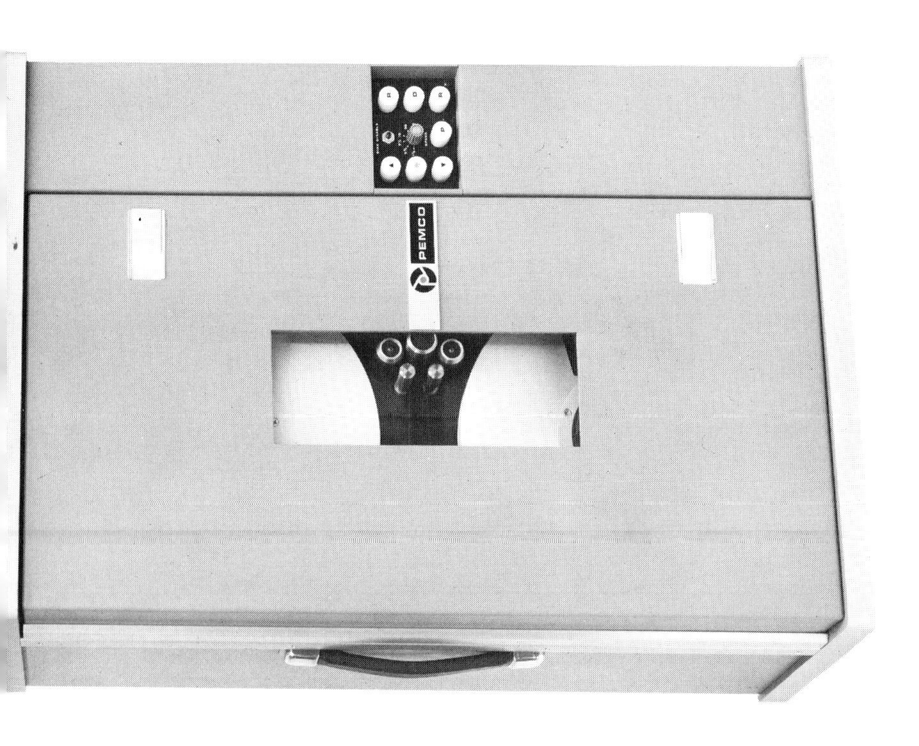

Microphone

Designers: Ampex Corp. Staff: Arden Farey, Industrial Design Division Mgr.; Frank T. Walsh, Director, Corporate Industrial Design Dept.

Intersecting cones produce shape of microphone, used as a table-top model or hand held. Strain relief on cord helps stabilize mike; die-cast zinc housing is heavier on bottom. Finished in clear epoxy on mike head and textured paint on base.

Microphone (bottom)

Designers: Ampex Corp. staff: Terrence Taylor, Sr. Industrial Designer, Arden Farey, Industrial Design Divisional Mgr.; Frank T. Walsh, Director, Corporate Industrial Design Dept.

Housing is precision zinc die-cast; head is diamond turned and clear epoxy coated, rotates freely to provide for changes in impedance.

Microphone (right)

Designers: Deschamps-Mills Associates: R. L. Deschamps, K. Wilson

Client: Shure Brothers, Inc.

Top of this two-piece assembly is an aluminum screw machine part; bottom is die-cast. Joint is emphasized with black Mylar nameplate. Black plastic threaded part holds stainless steel mesh grille against shoulder, and the tightening slots become decorative detail.

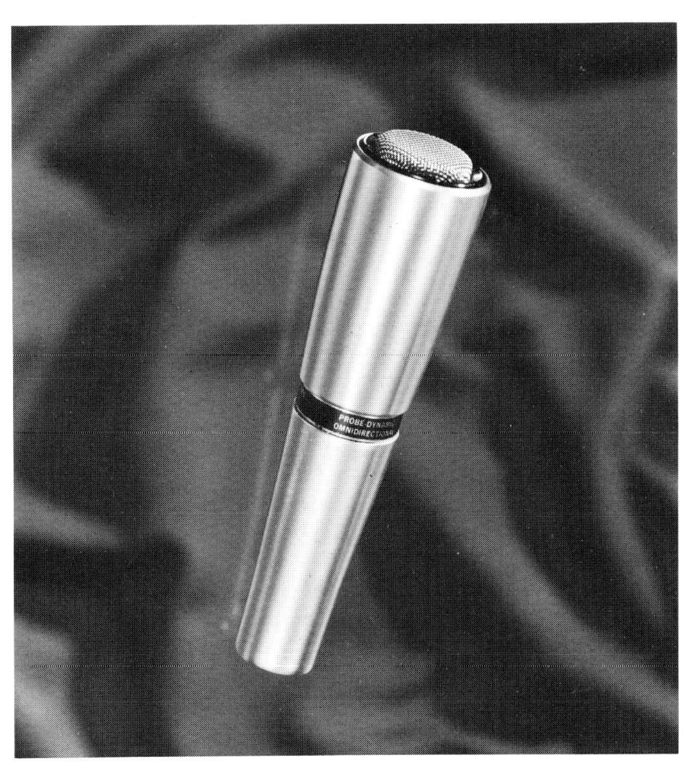

Helium-Neon Lasers

Designer: Clement Laboratories: Carl J. Clement
Client: Spectra-Physics, Inc.

A 5"-diameter heavy-walled tube connects mirror-mounting end plates. Hot plasma tube (which must be located between mirrors) is thermally isolated from main structure by sheet metal baffle, by-pass venting, and plastic foam insulation. Connections to external environment through three pins set into three spherical bearings assure that basic structure is independent of external stresses.

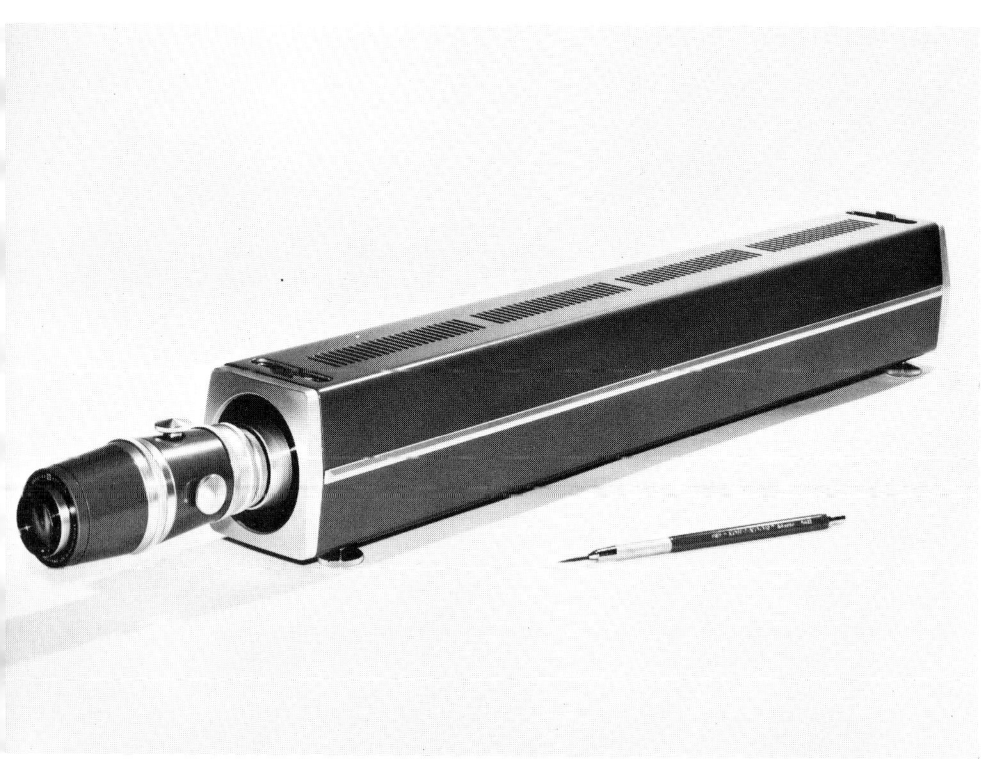

Spatial filter and lens family for lasers

Designer: Clement Laboratories: Carl J. Clement
Client: Spectra-Physics, Inc.

System for magnifying and cleaning laser beam requires an extremely stable and precise method of positioning apertures down to three microns in diameter. System must also accommodate various lenses and apertures interchangeably. Solution lies in design of "parent" spatial filter housing to accommodate all other elements. Although aperture is easily separable, its positioning stages remain with the housing, are controlled by radial knobs or coaxial ring.

Portable infra-red thermometer

Designers: Tepper + Steinhilber Associates, Inc.: J. Budd Steinhilber
Client: Raytek, Inc.

Instrument is used for contactless measurement of surface temperatures. Pistol-like form lends itself to aiming, triggering, meter reading. Infra-red sensing system, amplifier circuitry, and batteries are all contained within cast aluminum body. Barrel of satin-black anodized aluminum permits readings unaffected by temperature.

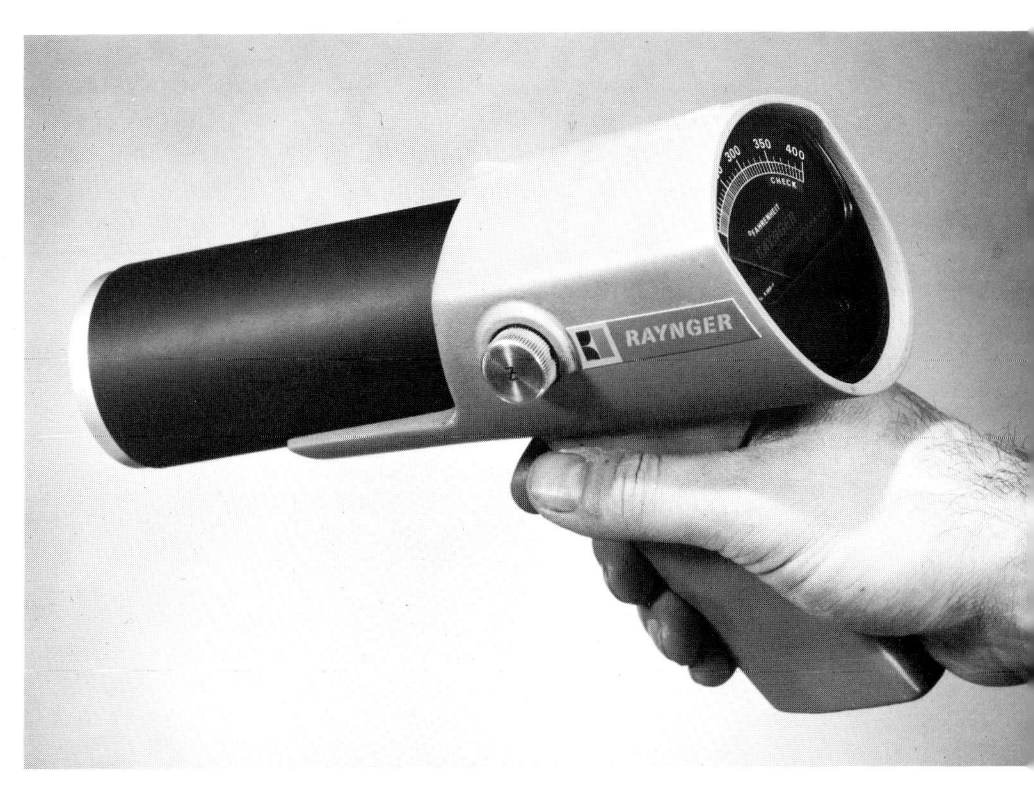

pH meter

Designers: Corning Glass Works staff:
C. Minot Dole, Jr.

Case of aluminum alloy for chemical and mechanical stability. Finish is baked dark gray epoxy; bezels, knobs and control plate of brushed aluminum with clear epoxy. Rear bracket supports input connectors, allows board to be self-supporting for assembly and repair.

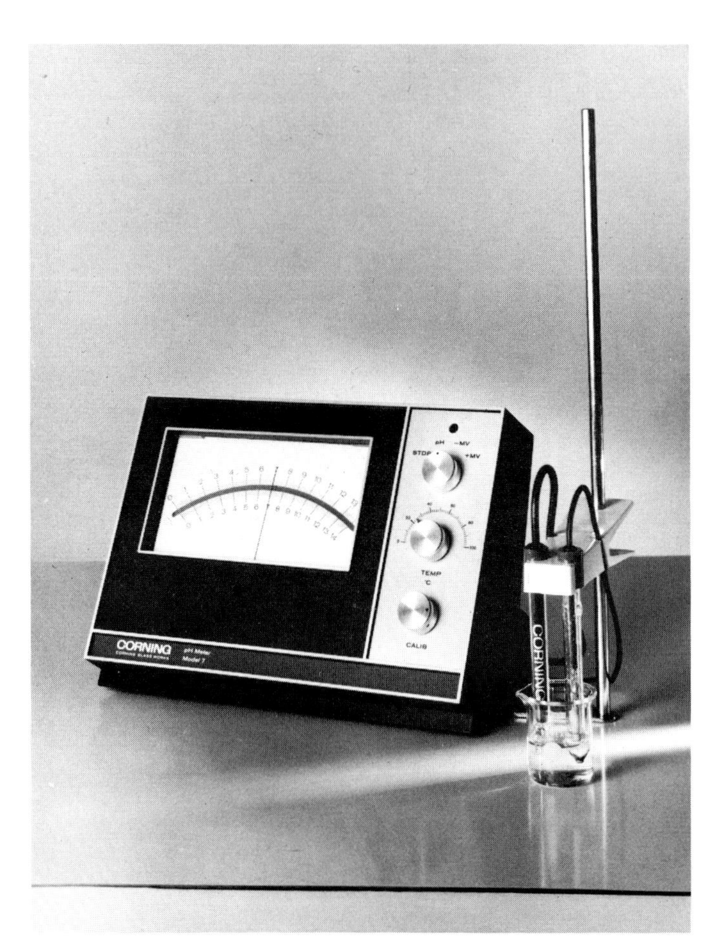

Portable pH meter

Designers: Knapp Design Associates, Inc.:
Courtney A. Leubner
Client: E. H. Sargent & Co.

One of three pH meters concurrently designed with as many common parts as possible. Sand-cast side pieces give line identity to all three models. Portable meter has brushed aluminum carrying handle; retractable legs tilt meter for readability.

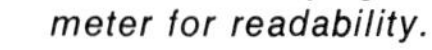

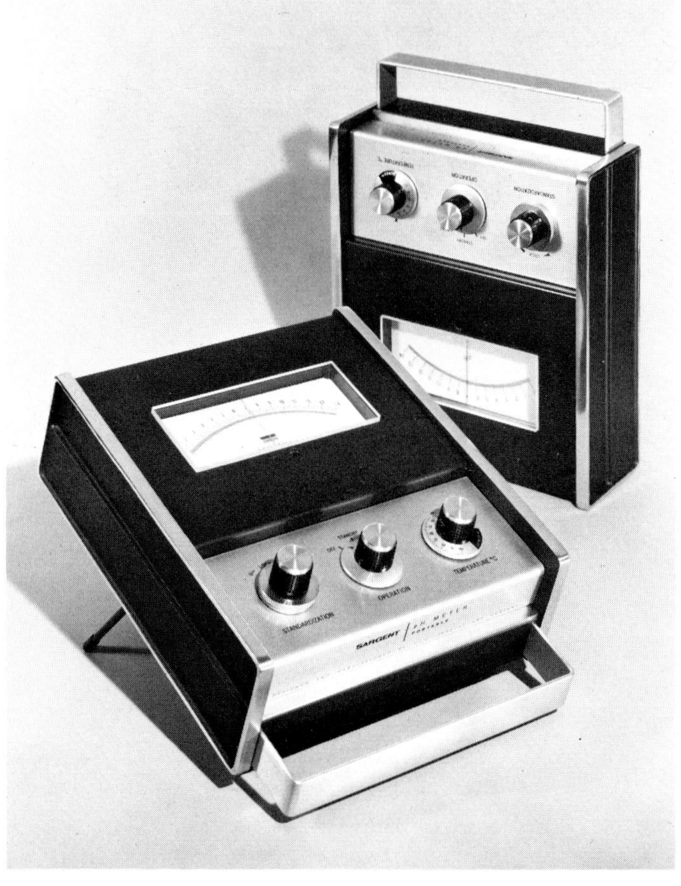

Panel meter

Designers: Yang/Gardner Associates, Inc.: Peter Quay Yang
Client: Apollo Manufacturing Co.

One-piece molded cover snaps on back-up plate.

Centrifuge balance

Designers: Corporate Design Consultants: Richard J. Koehan, William E. Sydlowski

Bottles in this laboratory instrument correspond to five rotor heads of varying radii, and balancing arm is an inverted trough designed to cradle the five bottles precisely the way they spin in rotor heads.

Student meter and beam balance

Designers: Stowe Myers Industrial Design: Stowe Myers
Client: Cenco Instruments Corp.

Part of a line of science laboratory equipment for schools and colleges.

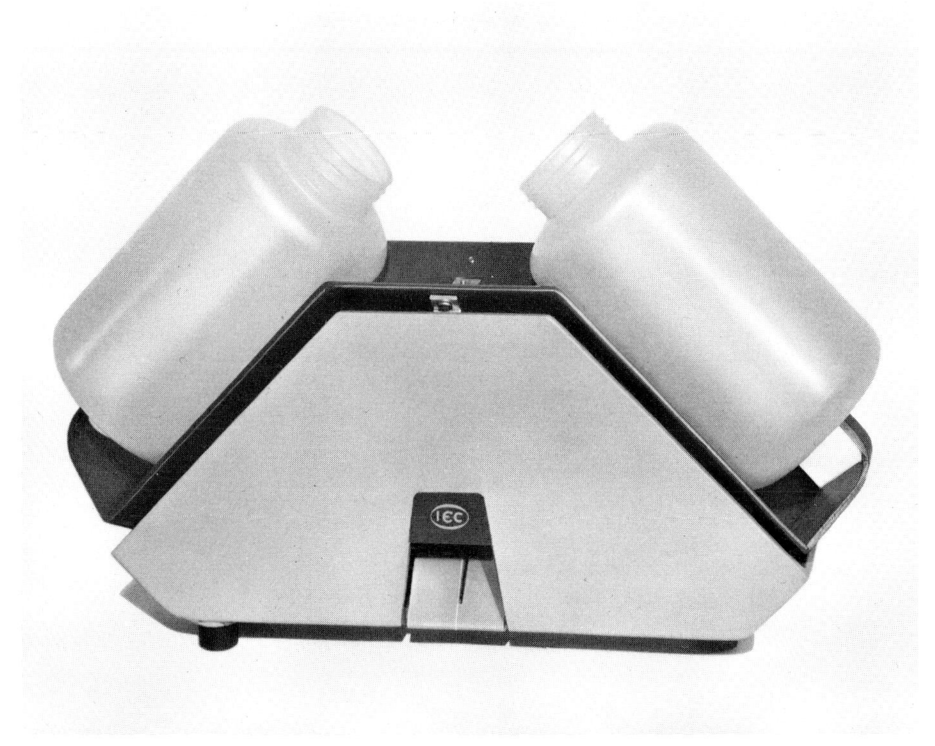

"Dispens-O-Disc" system

Designers: Ford & Earl Design Associates:
N. Kuypers, R. Grosso, C. C. Paul
Client: Difco Laboratories

Device used in biological and medical laboratories to dispense variety of reagent-saturated dehydrated discs onto bacterial test specimen. System consists of two basic components: (1) individual magazines, each containing 50 discs; (2) dispenser, which holds up to eight magazines.

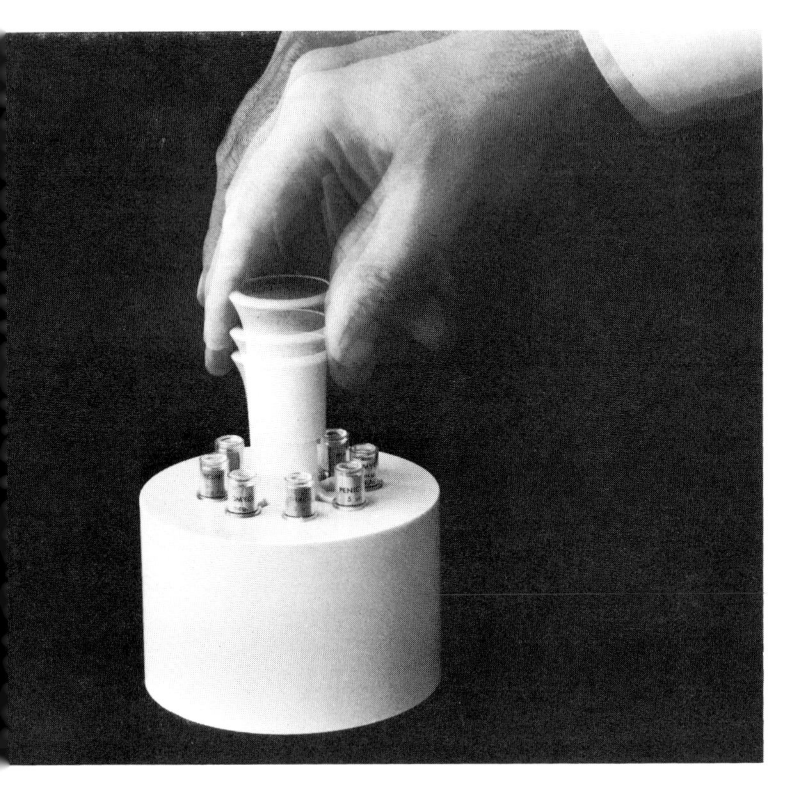

Physician's examination table and cabinets

Designers: Don Dailey Associates: Don Dailey & Fred Eilers
Client: Shampaine Industries, Inc.

Since a hidden mechanism raises table to correct height for examination, no step is needed for patient to get on or off. Welded steel frame, stainless steel and Formica facings, padded vinyl upholstery.

Disposable nurser and infant formula system

Designers: Don Dailey Associates: Don Dailey & Fred Eilers
Client: Mead Johnson & Co.

Filling equipment and safe-feeding device simplify and otherwise improve process of preparing and administering infant formulas in hospitals. Bottle and nipple washing, sterilization and refrigeration equipment are eliminated, space and personnel freed for other activities.

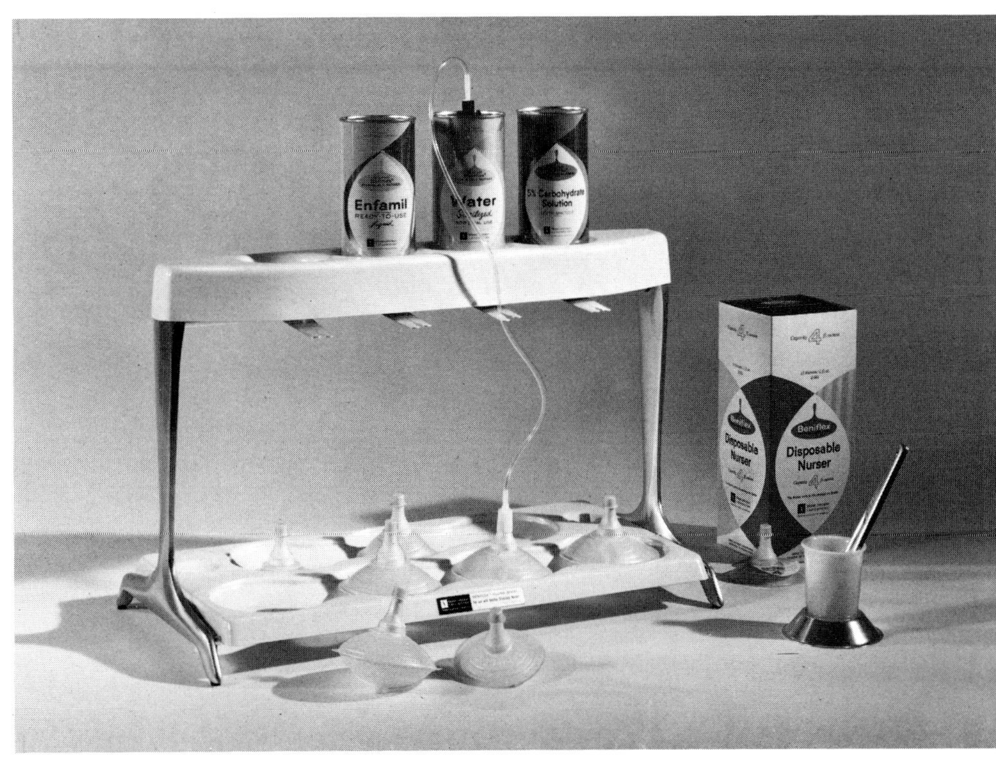

Blood pH system

Designers: Corning Glass Works staff:
C. Minot Dole, Jr.

Modular interior fabrication containing mechanical gear fits another expanded product system, for efficient inventory maintenance.

Laser photocoagulator

Designer: Morris Barnett
Client: Optics Technology, Inc.

Surgical instrument used in correcting retinal detachment. Millisecond burst of laser light focused on lens of patient's eye "welds" retina to choroid layer. Knob recessed to prevent accidental change of energy output.

Electrocardiograph

Designers: Hewlett Packard Staff: Carl J. Clement, design director; Indle King, Donald Pahl, Gerry Priestley, associates

Two bulky earlier models combined into one.

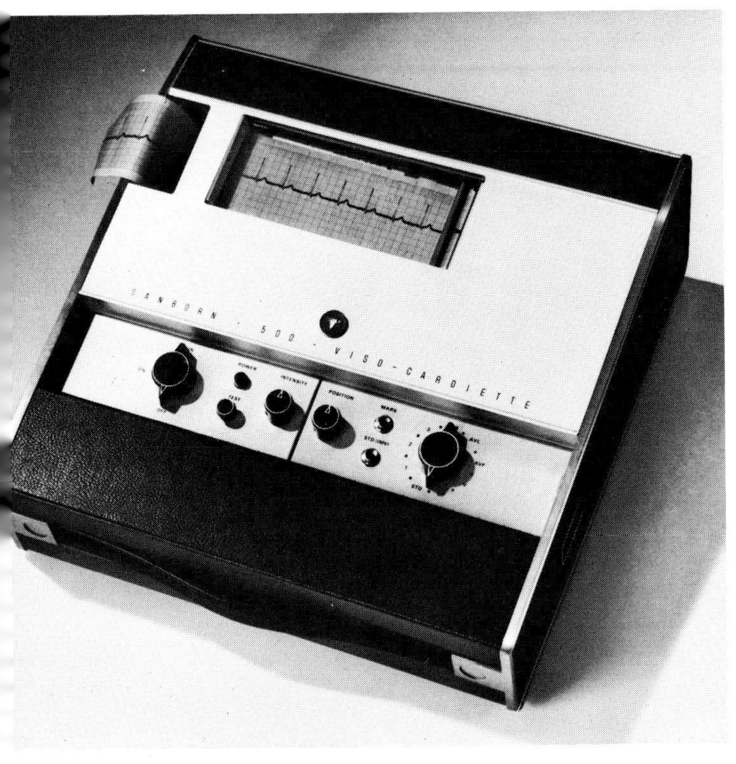

Infra-red radiometric microscope

Designer: John Bruce
Client: Barnes Engineering Co.

Microscope platform must be stable while allowing head to swing full 180°.

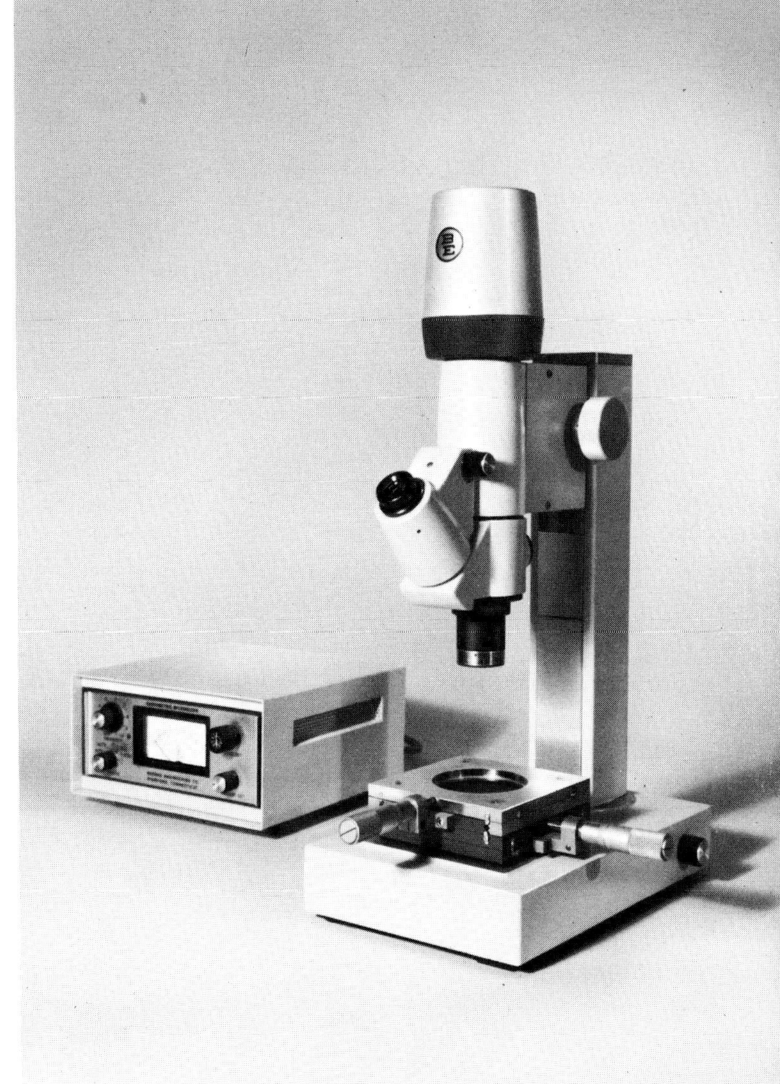

Tandem seating

Designer: Charles Eames
Client: Herman Miller Inc.

Chairs for public areas are pre-grouped in tandem. Vinyl cushions, identical for seat and back, are suspended on cast aluminum side members, can be replaced on location.

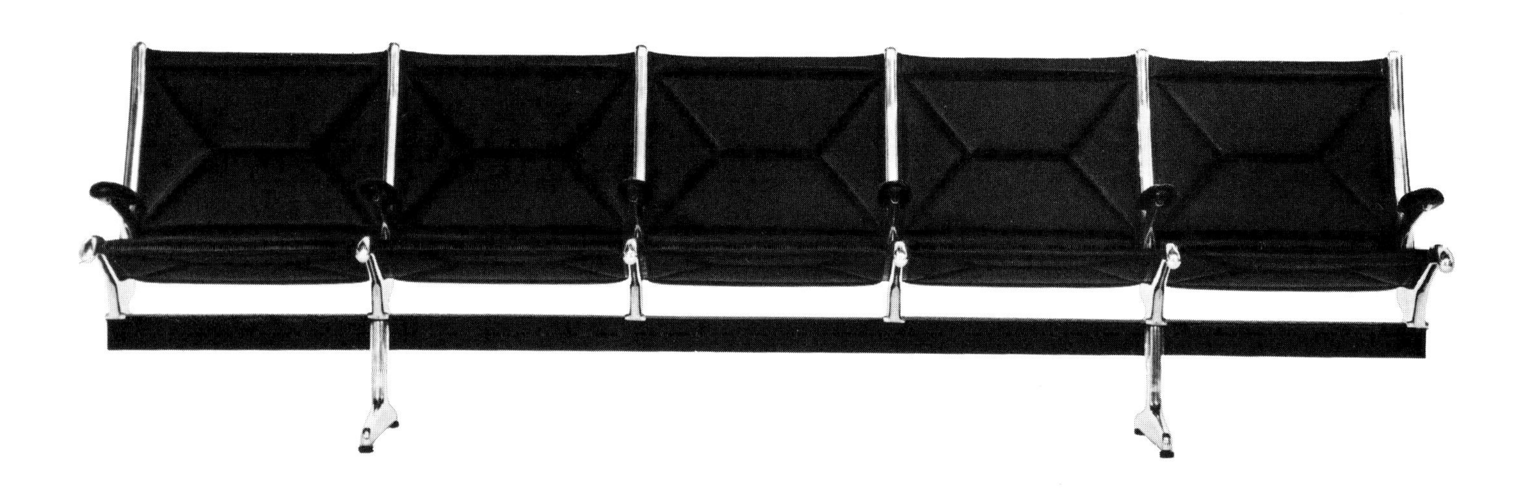

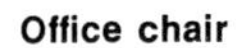

Barber's chair

Designers: Deschamps-Mills Associates: R.Massaccesi, R. Deschamps, C. Redfield
Client: Emil J. Paidar Co.

Design minimizes protruding parts that get in the way of the barber. Recline mechanism permits customer's head to be lowered to the lip of a shampoo bowl without requiring that he leave the formed plywood, vinyl covered chair. Metal band between seat and arms facilitates removal of hair that normally accumulates there. Leg rests, foot rests, back supports and pump handles are of chrome-plated castings.

Office chair

Designer: Leon Gordon Miller
Client: S. J. Campbell Co.

One of a modular group of stainless steel and wood cabinets, desks and chairs for office use.

Sink

Designers: American-Standard staff: Jack N. Kaiser; W. B. Winterbottom, Mgr. Industrial Design

Fittings are mounted in "pylon" above splash area, making possible a variety of fittings with minimal sink inventory.

Prefabricated lavatory

Designers: Williams Associates: W. J. Hannon, Jr.
Client: Market Forge Co.

Lavatory unit for hospital patient's room is prefabricated of extruded aluminum frame, stainless steel sink and shelf, wood grain vinyl clad steel skirt and panels. Can be recessed or wall mounted.

Panel lamp

Designer: Frank Klay
Client: Litecontrol Corp.

Concrete coffer-ceilings used in commercial buildings require compatible square-unit light sources that combine efficiency with visually low brightness. In this design, a GE 12" square panel lamp is combined with specially designed parabolic louver so that fixture appears unlighted at normal viewing angles. Uplight illuminates ceiling coffer itself.

Coffee vending machine

Designers: Labalme Associates: Douglas A. Long
Client: Rudd-Melikian, Inc.

Machine, designed for plant or office use, can be counter or wall mounted, push-button operated or coin controlled. Upper panel can be vinyl covered to match office decor.

Library furniture

Designers: Dave Chapman, Goldsmith & Yamasaki, Inc.: Kim Yamasaki, Emerson Purkapile
Client: Brunswick Corp.

Versatile line of furniture is based on suspension system that permits adjustable shelves, bases, tops, back and end panels, work tops and display props to be attached to the basic structure independent of each other.

School cabinets

Designers: Dave Chapman, Goldsmith & Yamasaki, Inc.: Kim Yamasaki, Robert LeSueur

Client: Brunswick School Equipment Division, Brunswick Corp.

Flexible, knock-down system of school cabinets used as room dividers, teaching surfaces, work counters, etc.

"Action Office"

Designers: George Nelson & Co., Inc., in collaboration with Robert Probst
Client: Herman Miller Inc.

Office system designed to work with or without walls, for individuals or groups, sitting or standing. Modular components provide visible storage, sometimes combined with work surfaces–e.g. file bin along edge of desk.

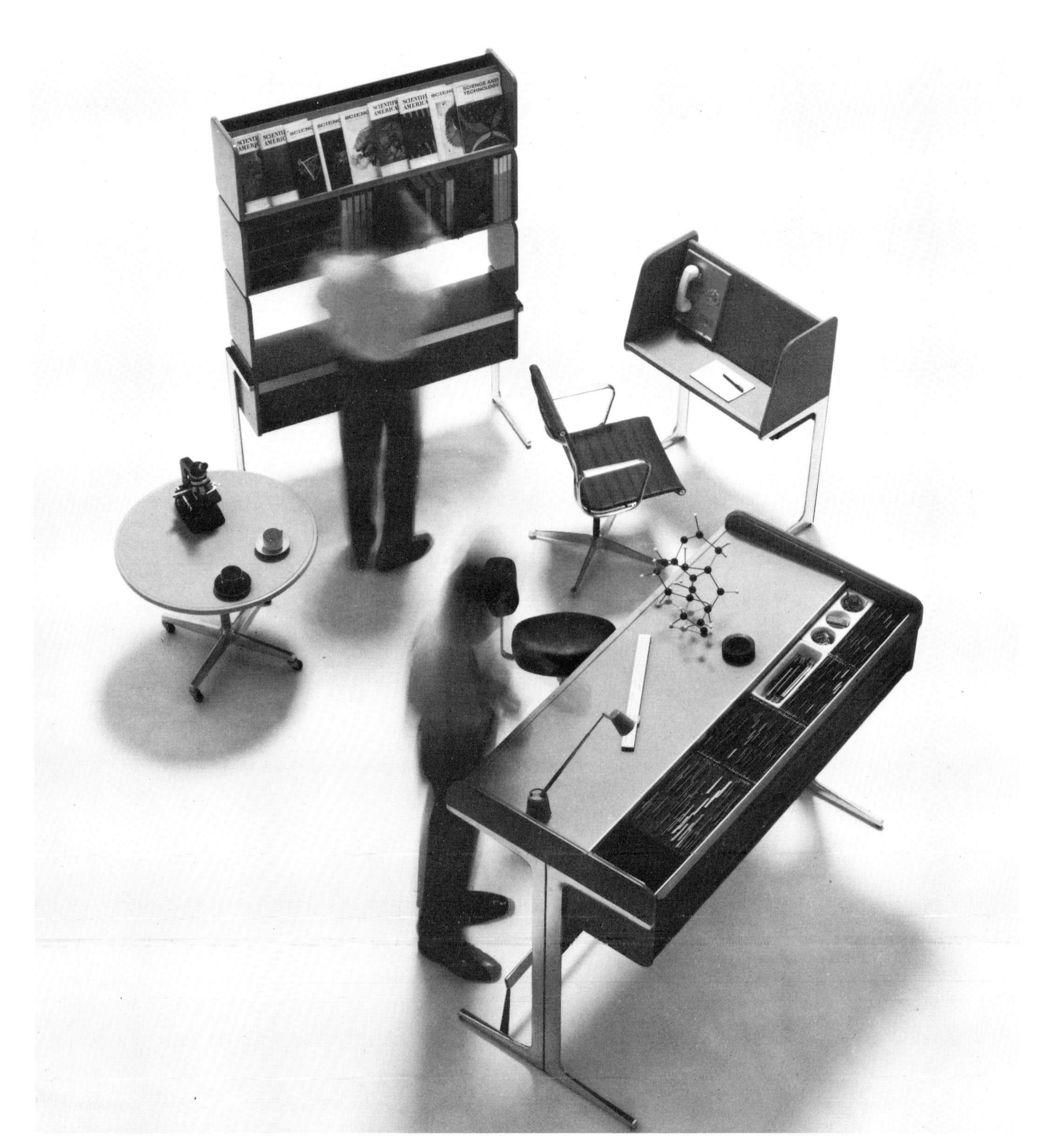

Stack chair

Designer: David Rowland
Client: The General Fireproofing Co.: Joseph H. McCarthy, Mgr. Product Development and Design

Chair stacks easily on rear supports of its 7/16-inch chrome-plated-rod frame. Specially designed dolly holds 40 chairs in 4-foot high stack. Contoured seat and back panels have textured vinyl finish or are upholstered in expendable vinyl or fabric.

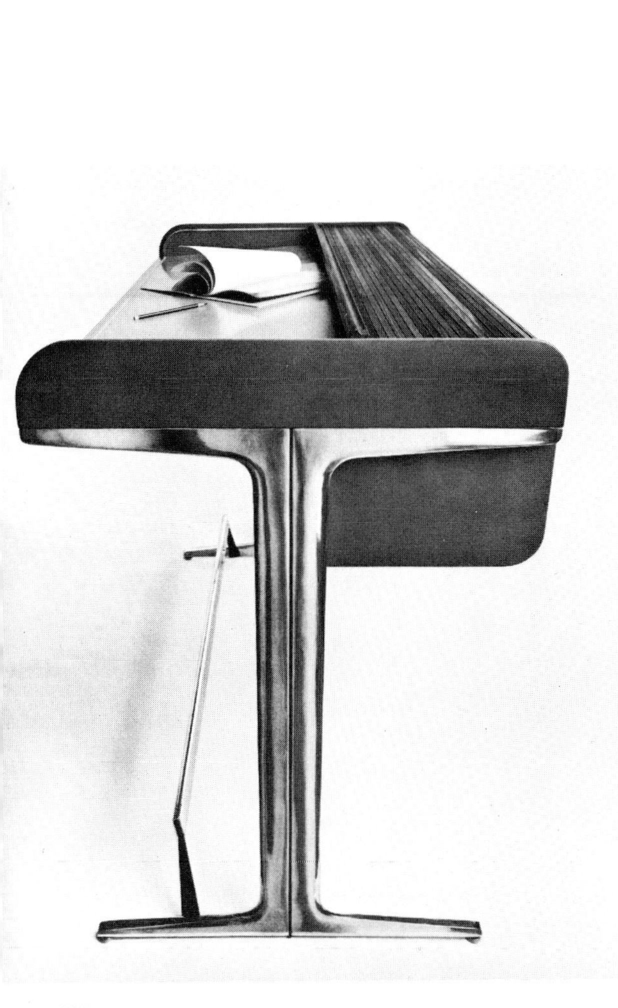

Electric office typewriter

Designers: Sundberg-Ferar: C. W. Sundberg, M. Ferar.
Client: IBM

Break in side elevation cuts down bulk.

Electric office typewriter

Designers: George Nelson & Co.: John Svezia, project director
Client: Olivetti Underwood

Two-piece clam shell enclosure with pantograph access to internal components. Carriage assembly has been lowered and aligned with front top cover to reduce size. Horizontal dual track, 3" diameter carbon spool permits convenient top loading without sacrifice of ribbon printing capacity.

Electrostatic copier
Designers: Armstrong/Balmer & Associates
Client: Xerox Corp.

Portable dictating machine
Designers: Eliot Noyes and Associates: Eliot Noyes, Ernest Bevilacqua, Allan McCroskery
Client: IBM
Full set of controls can be operated with one hand.

Dry photo-copier

Designers: Ford & Earl Design Associates: R. Grosso, S. Highberger, D. Saporito
Client: 3M Co.

Two aluminum die-castings house entire mechanical, electrical and lighting configuration. Top cover also serves as hinged platen to hold the original flat.

Dictating machine

Designers: Raymond Loewy/William Snaith, Inc.
Client: Dictaphone Corp.

Cover molded of Celcon acetal copolymer for dimensional stability.

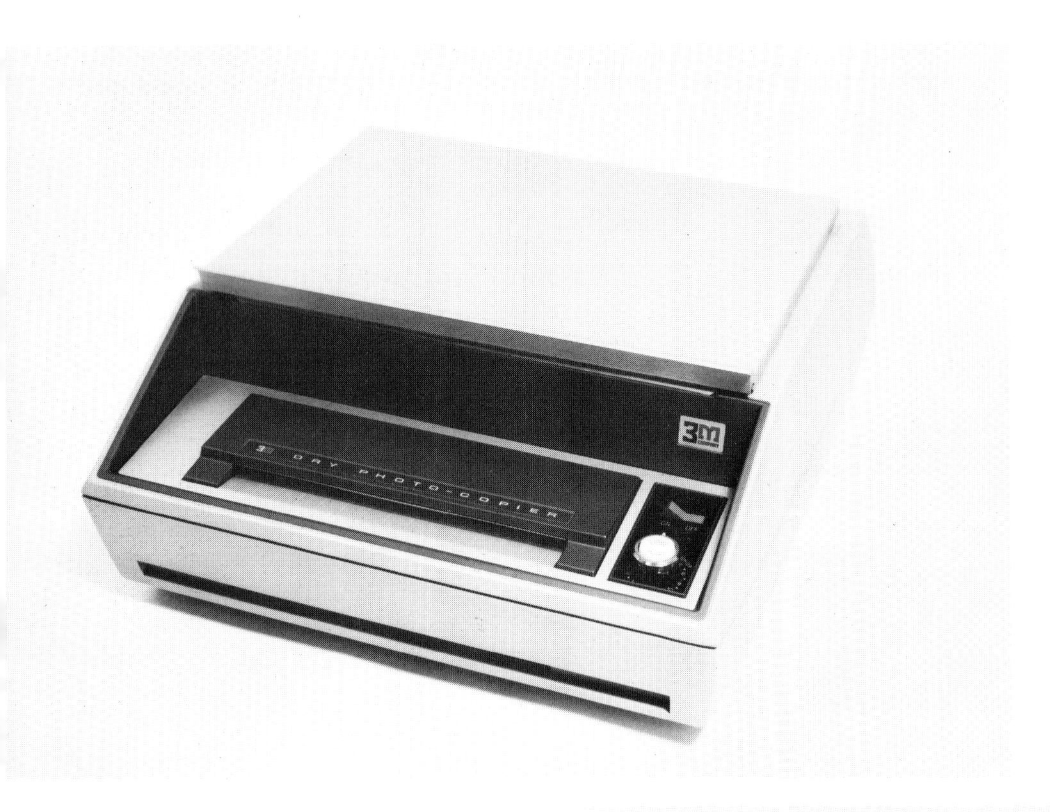

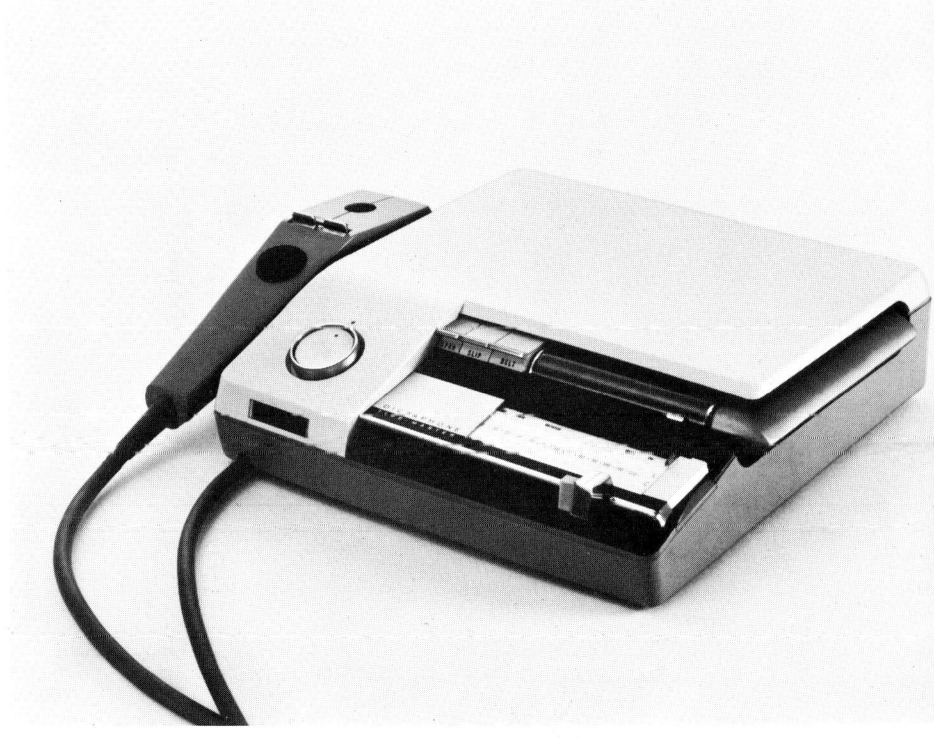

Electrostatic copier

Designers: Raymond Loewy/William Snaith, Inc.
Client: Pitney-Bowes, Inc.

Side covers come from common mold, are thus adaptable to right and left–reducing costs, simplifying inventory. Indicators at rear light up when fluid drops below required level.

"Telecopier"

Designers: McFarland/Latham Tyler Jensen, Inc.: Kalman Durik, Robert Arthur
Client: Magnavox Co.

Device (approximately the size of an electric typewriter) uses telephone system to transmit Xerox copies.

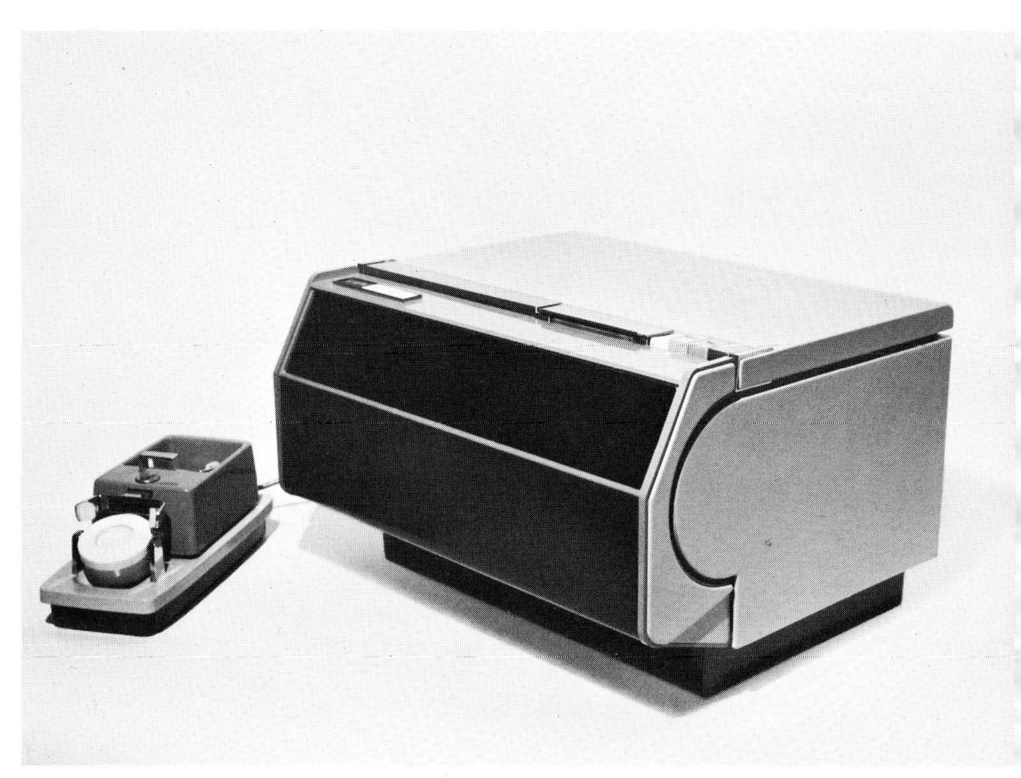

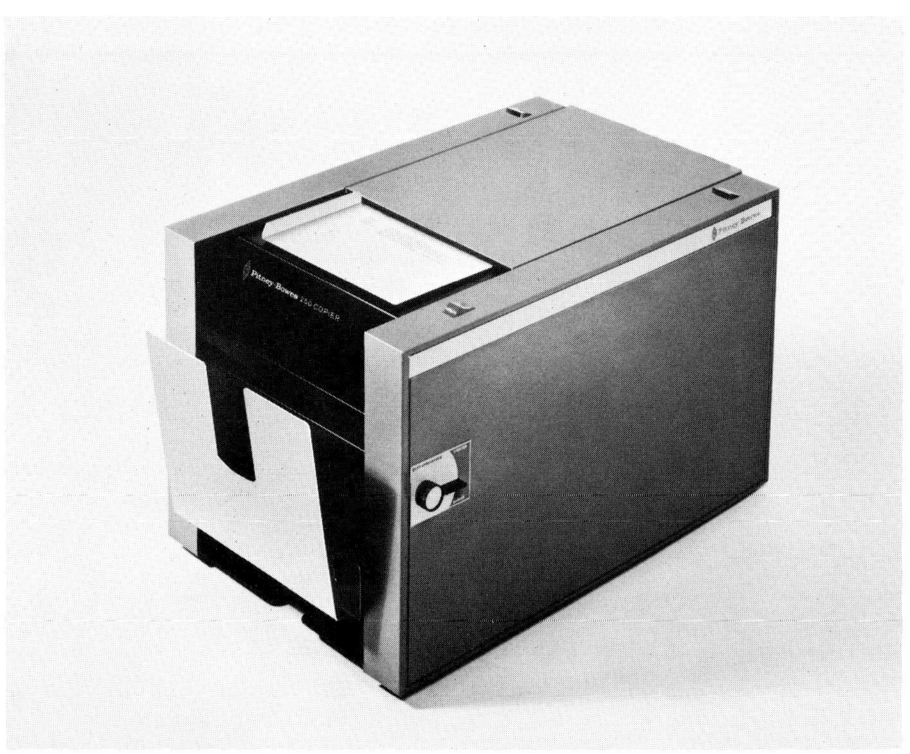

Postage meter

Designers: Friden, Inc. staff: Edward Salter

Lower casting serves as base, and is only part of unit requiring close tolerances. This makes possible "float design," which allows for casting mismatching.

Addresser-Printer

Designers: Raymond Loewy/William Snaith, Inc.
Client: Pitney-Bowes, Inc.

Cast iron, painted dark brown, is major design element. Light gray snap-on plastic covers conceal carriage support ends, match sheet-metal carriage cover.

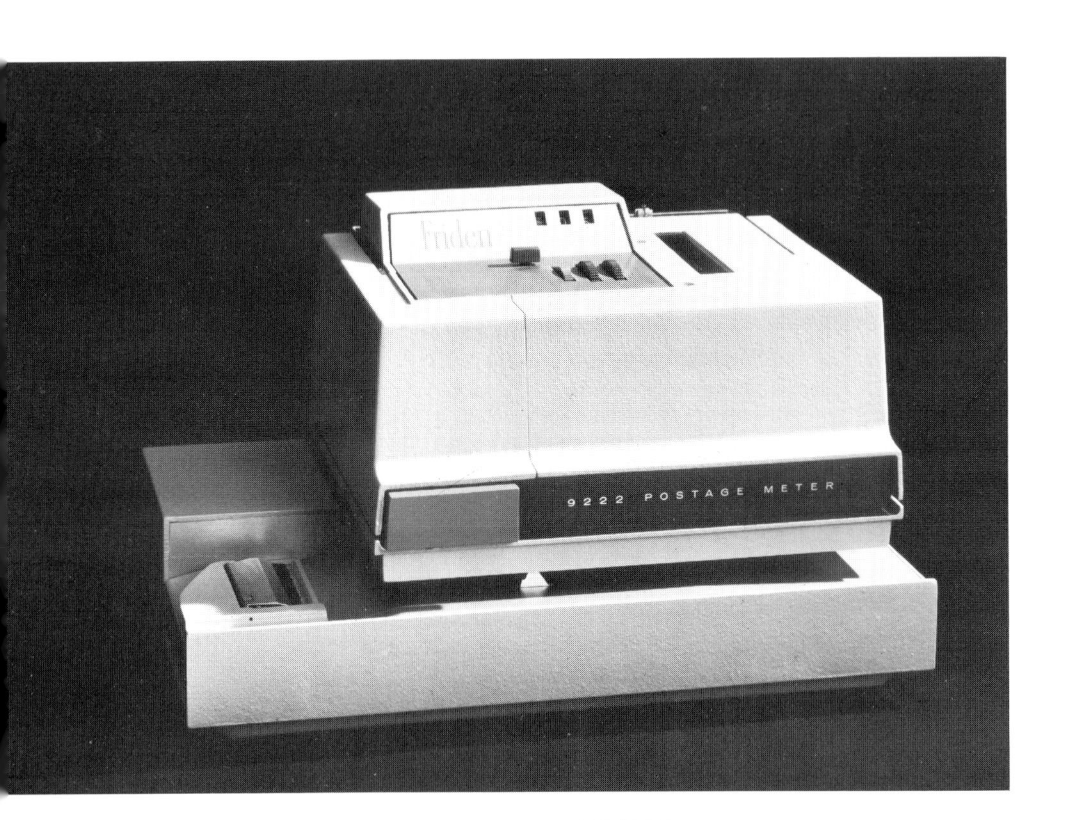

Addressograph

Designers: Addressograph-Multigraph Corp. staff: J. J. Van Acker, industrial design manager, C. F. Rudd, project engineer

Mylar ribbons, Nylon cams, vinyl-acrylic paint and aluminum were introduced to improve existing system.

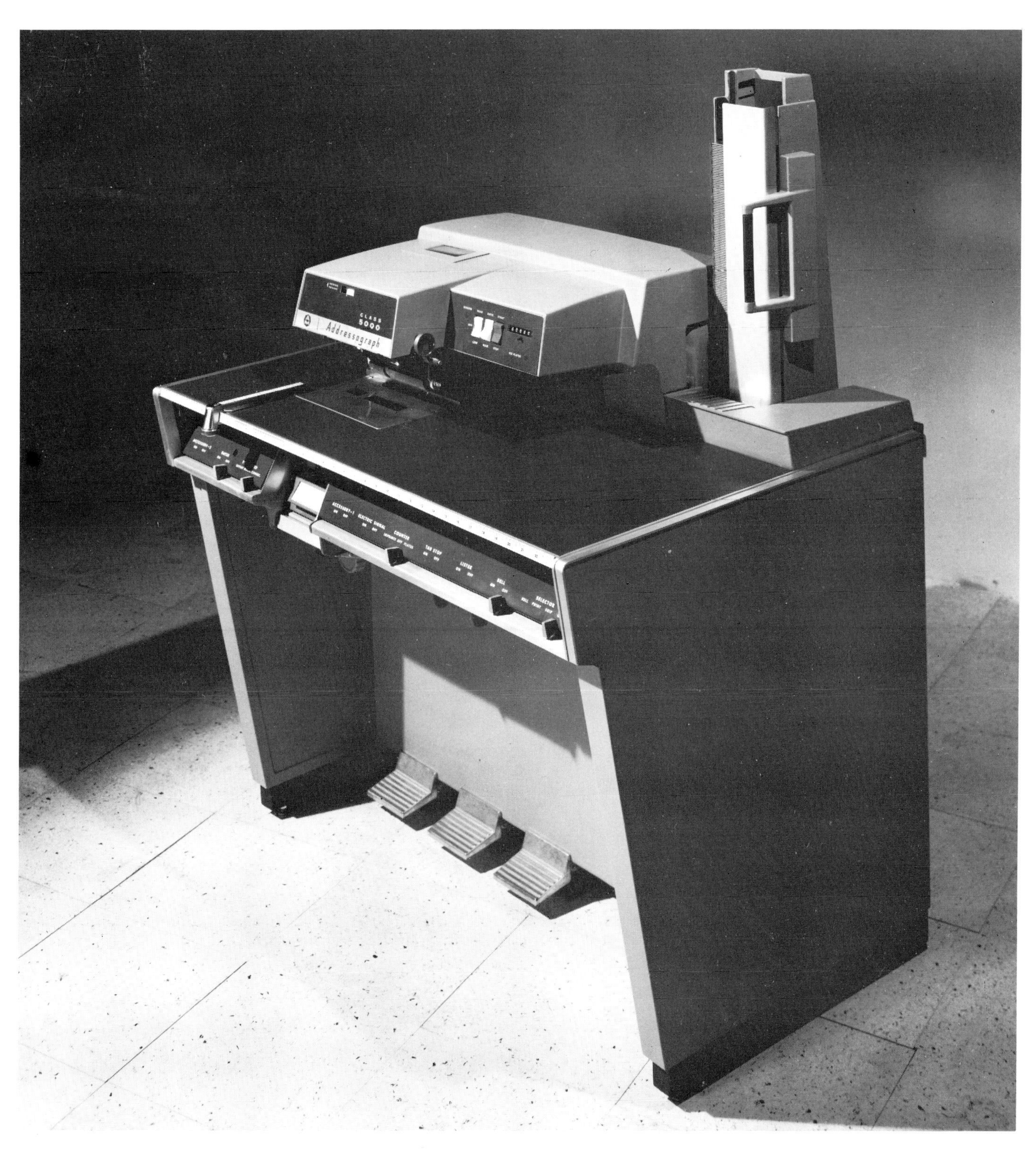

Part Two: Organizing Space

Organizing Space

Architecture is organization.
—Le Corbusier

Is R. Buckminster Fuller's geodesic dome (page 116) architecture or engineering? Does it really matter which it is, or whether it is either? It is one of the few designs in this collection that can seriously be said to make what young architects used to call a "statement." In its various manifestations it is a sermon, not in stone (so far) but in steel, aluminum, plastics, bamboo, and anything else appropriate to a given problem, and has made its point in applications ranging in function from railroad roundhouses to the U.S. pavillion at Expo '67.

Although industrial designers do not practice architecture as a rule, they design many of the components architects must specify, like the lavatory and kitchen sink on page 100. Also, industrial designers are increasingly called on to organize the limited, or special-purpose environment. Where a structure is a mechanism for performing a specific function, the industrial designer's experience is particularly relevant. It was always jarring to hear that a house was a "machine for living," particularly when that phrase was torn from all context. "Living" does not lend itself well to standards of machine performance. But it is not much of a shock to view a shoe store as a machine for selling shoes, or to see that service stations are mechanisms for attracting drivers, dispensing gas and oil, and selling tires, automotive accessories, rabbits-foot key rings, and soft drinks. Traditionally they are dirty and garish. They are also indispensable. Can they be more effectively organized? Can they sit more agreeably on the landscape? Some suggested answers appear on page 128 and 129.

The other environmental designs in this section are also mechanisms directed to a very specific aim.

Proposition: Water storage is necessary. Question: Is a water storage tank, therefore, a necessary *evil?* Answer (overleaf): Not necessarily.

Ever since arriving on earth mild-mannered Clark Kent has used telephone kiosks for his quick change into Superman. The doorless model on page 119 is spectacularly badly designed for his purposes, but it has some distinct advantages for the rest of us.

Motivation is what distinguishes the street environment on page 120. The designer initiated the project as a civic gesture, contributed his work, directed the construction, and got local firms and citizens to donate materials, labor, kibbutzing. His point was to beautify a tiny section of Akron, Ohio, and to demonstrate that tasks of this kind could be best handled locally.

An environmental design project of almost staggering size was the equipping of the U.S. Air Force Academy by Walter Dorwin Teague Associates, Incorporated. It entailed furnishing and planning some 3,500,000 square feet of floor space, for which the designers had to design or specify more than 1700 different items. While the most impressive feature of the project is its scope, there were unusual challenges in the details. Modern interiors had to be reconciled with the tradition of a spartan military life: all equipment had to be made to last 50 years, despite the hard use it would get from the young and lively cadets. The cadets room on page 128 had to provide storage and space to accommodate the several uniform changes cadets go through each day.

Geodesic Dome

Designer: R. Buckminster Fuller

One of many versions of designer's geodesic structures which, through balance of forces, provide maximum enclosure of space with minimum material.

Water storage tank

Designers: Cornelius Sampson & Associates: Cornelius Sampson
Client: East Bay Municipal Utility District

Parasol of six reinforced concrete arches rests on rim of standard water tank, continuing contour of the hill.

Transmission pole

Designers: Henry Dreyfuss Associates
Client: Southern California Edison Co.

Cantilevered insulators at the end of fixed-angle steel arms permit shorter wood pole than previously used.

Light pole

Designers: Donald Deskey Associates
Client: The City of New York

Extruded aluminum light pole with mill finish allows flexible adjustment of luminaires.

Telephone kiosk

Developed by Bell Telephone Laboratories–industrial design collaboration with Henry Dreyfuss & Associates

Doorless booth has accoustically treated interior walls. Back wall is curved glass; roof is translucent plastic to let daylight through and to permit ceiling light to serve as topside beacon at night.

Bankers Trust Building

Designers: Henry Dreyfuss Associates
Client: Bankers Trust Co.

Facade of pre-cast textured masonry floor-to-ceiling window frames contrasts with metal and glass sheaths of nearby buildings on Manhattan's Park Avenue. This is the first major building for which entire design responsibility—both exterior and interior—was given to an industrial designer.

Community beautification project

Designers: F. Eugene Smith Associates
Client: Inpost (an Akron, Ohio activities center)

Designer initiated project with idea of planting a tree, expanded it to include street furniture, landscaping, signs, wastebaskets, ash trays and decorative lighting.

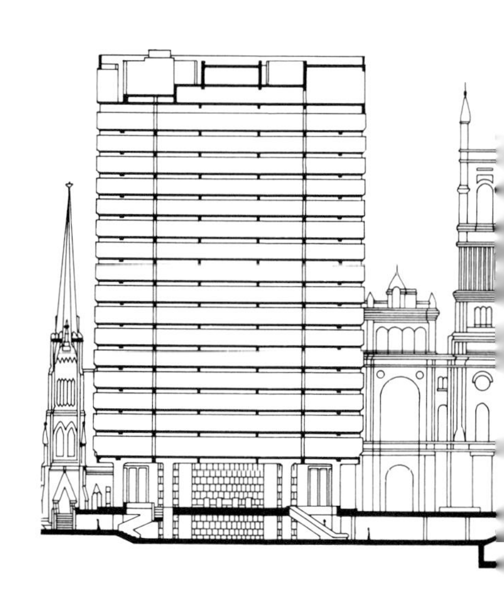

City Hall annex

Space planners: Becker and Becker Associates, Inc.: Nathaniel Becker, Maurice Robbins, Herbert Newmark.
Architect: Vincent Kling
Client: City of Philadelphia

The industrial designers gathered and analyzed data, established functional space requirements, collaborated with architect in developing interior spaces.

Lobby

Designers: Ford & Earl Design Associates: D. Wahler, W. B. Ford II.
Architect: Minoru Yamasaki
Client: Reynolds Aluminum Co.

Main lobby is three stories high, topped with skylight of aluminum and glass. Interior solution was to design an aluminum reception desk and place it–with planter areas and seating group–on a royal purple carpet.

Company president's office

Designers: Ford & Earl Design Associates: R. Adams, G. Benkert, W. B. Ford II

Space is divided into (1) working center with large table-desk, (2) informal seating area. Lighting is luminous ceiling membrane combined with downlights.

"Creative Marketing Center"

Designers: Becker and Becker Associates, Inc.: Nathaniel Becker, Frederick Leigh, Eugene Klumb

Client: International Paper Co.

Facility is used for client presentations and market testing. "Presentation Area" is equipped with wide range of audio-visual resources, and affords view of "Environmental Area," which serves as simulated retail operation or as display. "Audio-Visual Information Center" uses sound and rearview projection to describe client's products and services.

Communications seminar center

Designers: Becker and Becker Associates, Inc.: Nathaniel Becker, Frederick Leigh, Granville Ackerman
Client: American Telephone & Telegraph Co.

The client regularly holds one-day communications seminars for invitational audiences of 25 top-level corporate executives. The center—especially designed for these seminars—consists of a presentation theatre, lounge facilities, library, reception area and staff offices. The theatre's a/v capabilities include pre-planned lighting and single or multiple image projection on 34′-wide screen wall.

Board of Directors room and garden

Designers: Ford & Earl Design Associates:
D. Wahler, W. B. Ford II
Client: Bundy Corp.

Since board room, on top floor of old building, has poor exterior view, an inside garden was created, with the room set in its center like a box with two glass sides.

Bank interior

Designers: Walter Dorwin Teague Associates, Incorporated
Client: The Bank of Bermuda Ltd.

Lighting is used to distinguish public banking area from officer's "island." Banking hall design is intended to achieve "cool, airy" effect.

Office work areas

Designers: Ford & Earl Design Associates: D. Pierson, W. B. Ford II
Client: First Federal Savings & Loan Association of Detroit

After studies of traffic flow, materials and paper handling, projected growth, etc., total interior design program was formulated.

Employee cafeteria (and lounge)

Designers: Ford & Earl Design Associates: C. Benkert, W. B. Ford II
Client: Northwestern National Life Insurance Co.

Aim is to make lounge and cafeteria a complete change from work environment, by use of high ceilings, arched windows, tall trees, bright colors, warm woods.

Boeing interior

Designers: Walter Dorwin Teague Associates, Incorporated
Client: The Boeing Co.

Wood paneling, low furnishings are used to create "relaxed atmosphere" that is "warm but businesslike." Open staircase contributes feeling of spaciousness.

Dormitory room

Designers: Walter Dorwin Teague Associates, Incorporated
Client: U.S. Air Force Academy

Typical two-cadet room is heated by metal "wafers" set under windows. Curtains are of specially designed Dacron, resistant to dust, moisture, sun.

"Pump island complex"

Designers: Lippincott & Margulies, Inc.
Client: Esso

Lighting fixture, "merchandiser," and utility desk are designed to focus attention to busiest part of station.

Service station

Designers: Eliot Noyes and Associates: Eliot Noyes, Arthur DeSalvo, H. B. VerBryck
Client: Mobil Oil Corp.

Designed to enhance both efficiency and appearance of filling station.

Service station system

Designers: Peter Muller-Munk Associates, Inc.: Glenn W. Monigle, Raymond A. Smith
Client: Standard Oil of Ohio

Modular steel system effects flexibility in size and in plot orientation.

Part Three:
Image and Expression

Image and Expression

Design is communication.
—Francois Dallegret

Amen.
—Richard S. Latham

At least design *includes* communication. By looking like what it is and does, an object helps tell someone what it is and does; and it may tell far more. Much design, though, "is" communication in that it has no function other than to deliver a message.

The immediate purpose of such design may be to say something about a particular product; a long-range purpose may be to say something about the company that makes the product. In either case the message is probably one that can't be delivered verbally in the same time or space, or that doesn't lend itself to verbalization for other reasons.

A corporate design program is a major part of a company's effort to control the way it comes across. Corporate symbols, trademarks, exhibits, advertisements, packages, point-of-purchase displays are all agents of control. As agents they are like wartime spies in Casablanca: their job is made difficult by the presence of so many others in the same business. The American retina the country over is assaulted by visual directives to buy, rent, eat, smoke, drink something, or just fall in love with a corporation. Design is a competitive strategy for getting a message through the visual jungle. If it is humane design it will help civilize the jungle, too.

Exhibiting

The design of exhibits has increased with the political importance of international expositions, the burgeoning trade fairs, the increasing number of conventions, the enormous amount of exhibit space in commercial building lobbies, airports, etc. The concept of exhibiting has shifted towards the dynamic: designers try to make an exhibit less a matter of walking past a display of information than of participating in it. The visitor is no longer just a viewer, but a listener as well, and a *toucher.* He pushes buttons, lifts phones, operates machinery.

The designer of an exhibit usually has a much more direct and influential role in determining *what* is to be communicated than he has in other forms of communicative design. The content and form of an exhibit are sometimes inseparable, and by his choice of photographs, mechanisms, interactive devices, motion pictures, etc., the designer shapes what is being said.

So the first element of exhibit design is defining the content, trying to isolate the statement or information that is to be communicated, and trying to decide whether it can be communicated to the desired audience by means of an exhibit. Many exhibits go from the concept into a storyline, a point by point description of what the visitor is to see, feel, experience. Decisions about structure and materials grow out of the storyline.

While the physical design of an exhibit must be solved in terms of materials and structure, the root problem is always the handling of people. It is very difficult to determine how an exhibit will actually be used. But the designer can determine the optimum way for his purposes and try, by force of content and by traffic patterns, to encourage as many people as possible to go that route. Beyond that, he will provide options for visitors: the option of skipping certain parts, the option of starting in any of several places, the option of choosing their own pace.

That is one reason (although by no means the only one) that exhibits in recent years have tended to rely heavily on (and in some cases consist of) films. A motion picture happens in a fixed period of time, and usually in the dark. The audience chooses whether or not to enter the tent, but once in, it is pretty much a captive audience. (A circumstance carried to its logical extreme at the IBM pavillion in the 1964 New York Worlds Fair, when the audience took seats along a grandstand-like "people wall," which rose on a hydraulic elevator up into a theatre enclosure 54 feet in the air, and stayed there until the film was over).

Exhibit

Designers: Fogleman Associates: W. C. McDade
Client: Interchem Printing Inks

Designer sought to create "oasis" within clutter of National Packaging Exposition. Each conference room had its own mural and was partially closed by 5′ module panel.

Exhibit

Designers: Van Dyck Associates
Client: Pratt & Whitney Aircraft, division of United Aircraft

Multi-colored panels of old flight engravings provide contrasting background for display of jet engines.

Exhibit

Designers: Raymond Spilman Industrial Design: Raymond Spilman
Client: Office of International Trade Promotion

Water use in arid Peru was theme of International Trade Fair in Lima. The U.S. symbol used windmill with red, white and blue blades.

Exhibit

Designers: Michael Lax and Associates: Michael Lax
Client: Monsanto Textile Division

Photographs and sketches representing the ideas of eight designers are displayed on eight towers ranging in height from 10′ to 20′ Towers are mounted on, and illuminated by, glass platforms containing supplementary graphic material.

Storyboard for a film

Designers: Elaine and Saul Bass
Client: Eastman Kodak Co.

"The Searching Eye," a film presented at the New York World's Fair, employs a wide variety of filmic techniques to show the discoveries of the human eye.

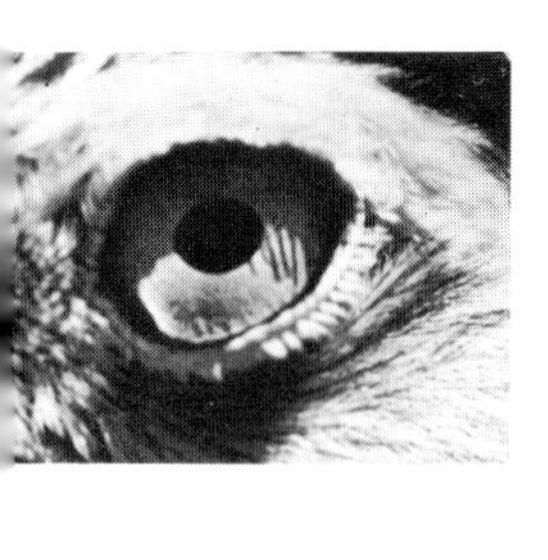
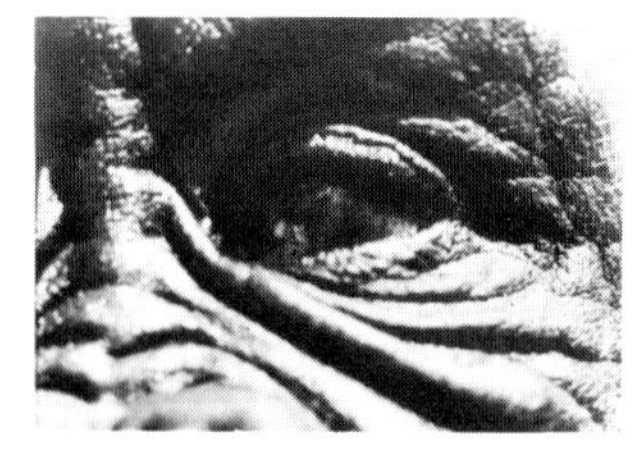

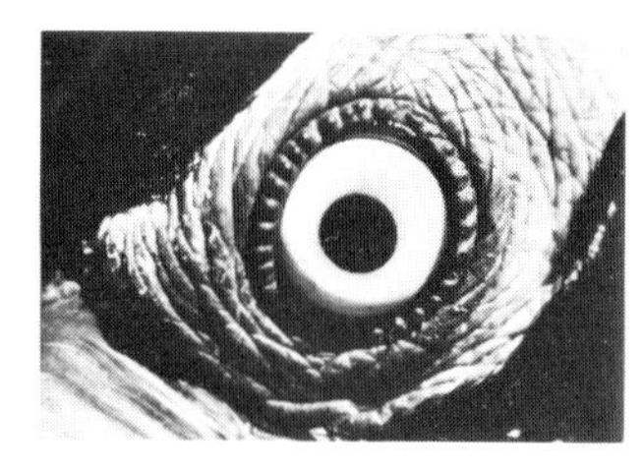
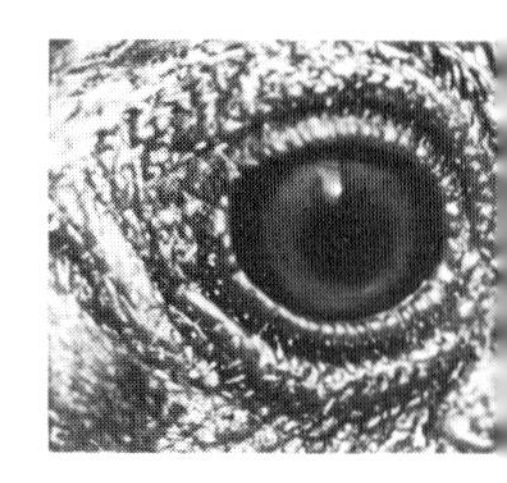
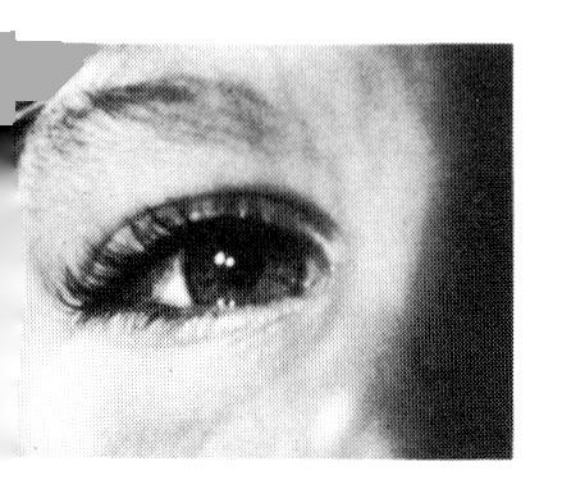

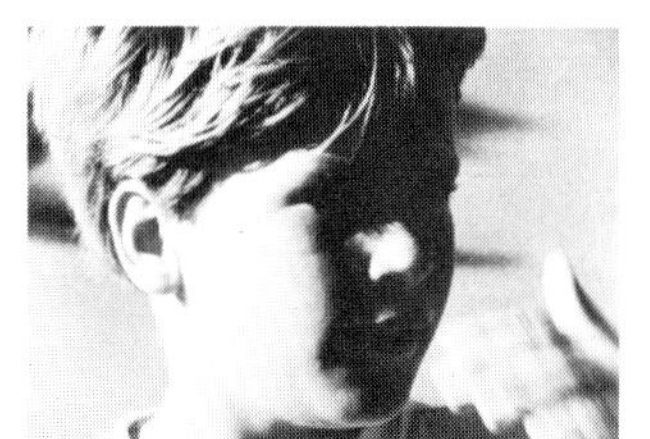

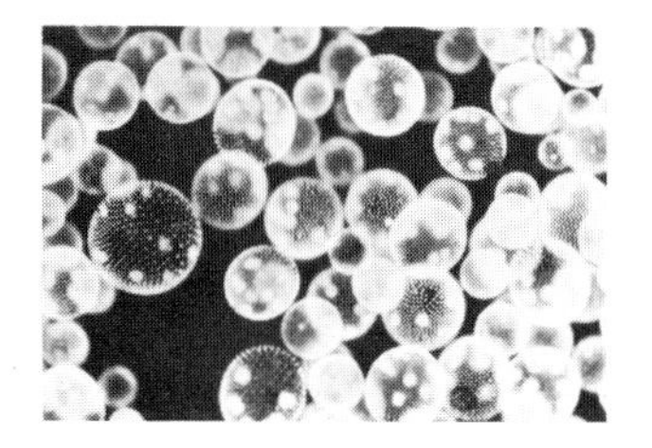
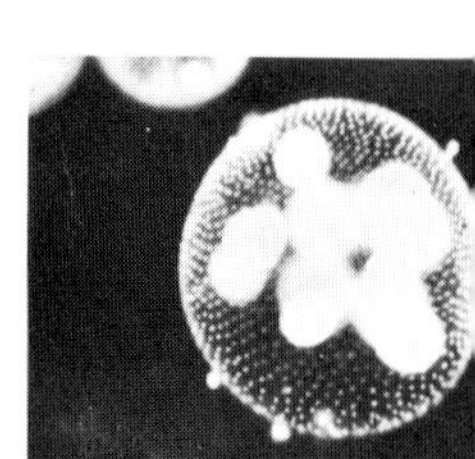

Industrial film

Designers: Morton Goldsholl Design Associates, Inc.: Morton Goldsholl, James Miho, Tom Freese
Client: Champion Papers, Inc. (sponsor); Needham, Harper & Steers, Inc. (agency)

Film tribute to American railroads deals with trains and the work and lives of people who run them.

Signs and Symbols

On the letterheads of many American design offices, and in the practices of many others, "corporate identity programs" have a prominent place. The importance of this service reflects the proliferation of companies, not always competitive, that are enough alike to be indistinguishable from each other. The confusion is compounded by the complexity of modern business, particularly big conglomerate business.

Consider the Indiana Pipe Wrench Corporation, which still manufactures pipe wrenches, but also makes dozens of products as unlike each other (and as unlike pipe wrenches) as electric shavers and medical instruments. The firm decides, perhaps in collaboration with a designer, to change its name to Hoosier Industries. A massive advertising and promotional campaign is launched, telling the world of the change. (Will the world pay attention? The designer is one of many people responsible for seeing that is does.)

For the new name, a new logo must be designed—one that will express the character of the company, particularly its place in the vanguard of technology. Because of the breadth and diversity of the Hoosier operation, the logo will appear in a great many different kinds of applications: letterheads and business cards and checks, architecture and company planes, delivery trucks and uniforms and advertisements. It will also be used on "literature." (The old Indiana Pipe Wrench Company used to pack a leaflet in with its pipe wrenches, congratulating the buyer and telling him something about wrench maintenance. But the new Hoosier Industries finds itself in publishing in a big way. It and its advertising and publicity agencies regularly issue a cascade of reports, brochures, recruiting pamphlets, instruction manuals and house organs—and the new logo will figure prominently in each of them.) The logo will have to function effectively both alone and in tandem with the logos of the corporation's several subsidiaries, and with the trademarks of the products they make.

Since mistakes in this area can be costly, no company seeks to alter its image recklessly. And, for all the time it takes, a corporate identity program is never done once and for all. It is a continuing endeavor that must allow for the expansion of services, the development of new products, and the extrusion of more literature. When a corporation, like Hoosier, is far-flung, with affiliates and subsidiaries using their own advertising agencies and other services, a designer rarely is in a position to supervise the application of the identification program. How can he be sure that his design will not be cheerfully subverted by a printer in Crawfordsville, Indiana, who likes the logo in magenta, over-printed on a line cut of a pipe wrench and accompanied by Old English script?

An instrument for aiding a company in the consistent application of a design, and for preventing its subversion, is the graphic control manual—a comprehensive guidebook to all the ways in which the graphic elements of a given identification program may be used. It would hardly be possible to maintain an identity program of any complexity without such a device, and the development of the control manual is in itself a large design undertaking.

It is tempting for designers to make a mystique of corporate identity, and tempting for clients to hope that a new graphic treatment will amount to a new lease on corporate life. It won't; although it can help sustain one. A company's identity depends more on the quality of its products and services and people than on the quality of the typography in its annual report. Many corporate identity programs are shortchanged whenever the consumer comes in contact with any aspect of the company more personal than a trademark. No amount of color matching can make up for inadequate quality control, and the friendliest of symbols cannot give a lasting friendly image to a bank whose tellers are rude. But by bringing all presentational aspects of a company into harmony, a good design program—in this case the consistent application of intelligence, experience, and professional skill—can increase and enhance corporate visibility.

In corporate symbols, as with any design, there is seldom any single demonstrably right solution. More than other designs, however, good symbols are likely, in time, to seem inevitable. There are of course parameters. It is unlikely that the same mark could effectively serve both a boutique and a manufacturer of heavy farm equipment. But there is a great deal more latitude than was once suspected, and companies wear brandmarks and colors today that they would have resisted (and did in fact resist) in other periods.

For a company almost inextricably established in the public mind to take on a new logo is an act that has far-reaching ramifications. Moreover, it is necessarily an act performed in public. For IBM Paul Rand took the corporation's existing logo and redesigned it to look contemporary and advanced, as the corporation is. The change was largely a matter of improving proportion and clarity. But for Westinghouse, which offered a different problem, the same designer created a wholly new logo. It did not, upon introduction, look the way Westinghouse ever *had* looked. Yet it is effective in the legions of uses to which it is now put.

The two bank symbols on page 156 illustrate how far apart in character two symbols for roughly the same kind of organization can be. The Marine Midland system consists of more than 200 branch banks, which were using a wide variety of typographic styles and several versions of the bank's original symbol. The effect was a fragmented identity. After their identification study established the desirability of a new symbol, the designers made one, retaining the ship for continued recognition through the changeover period. The design is reinforced by a manual to control the use of the symbol, color, type face, architectural applications.

The Marine Midland symbol is representational, almost literal. So is the symbol of the Savings Bank Association, which uses a flourishing tree and coin shapes in a mark intended to distinguish savings banks from other banks and from savings and loan associations. The symbol of the Chase Manhattan, however, is *non*-representational. One could argue that it has a general feeling of solidity, and that the related masses connote trust and stability. (At any rate it isn't a fragile symbol that connotes shakiness). But its success lies not in any specific connotation, but in the way the giant bank uses it in its many variations. The symbol says "Chase Manhattan" because the company and the designers decided that it would, and because the public—encouraged by a sizable and steady advertising campaign—quickly agreed.

There are of course more limited and precise needs, and marks to fill them. The visual pun on page 151 ("Anatomy of a Murder") was used for the promotion of a single motion picture. The Information International symbol on page 154 has something very special and limited to say and a relatively special group of people to say it to. And the mark for the Foulds Macaroni Company—a sheaf of wheat windblown into a spaghetti-like limp *F* —pleasantly connotes the maker and his product.

Exterior graphic treatment
Designers: Lippincott & Margulies, Inc.
Client: Eastern Air Lines

Exterior graphics and paint scheme
Designers: Raymond Loewy/William Snaith, Inc.
Client: Northeast Airlines
Curving air-foil graphic form is in contrast to most aircraft paint treatment, which is determined by window line. Color scheme is sunny yellow and white, with black lettering and logo.

Corporate symbol
Designer: Paul Rand
Client: Westinghouse Electric Corp.

Corporate trademark
Designers: Walter Landor & Associates
Client: Royal Industries

Logotype
Designer: Paul Rand
Client: IBM

Plant sign
Designers: Lester Beall, Inc.: Lester Beall, Richard Rogers
Client: Titeflex

Sign
Designers: Goldsholl & Associates: Morton Goldsholl
Client: International Minerals & Chemical Corp.

Corporate identity program
Designers: Latham Tyler Jensen, Inc.
Client: Archer Daniels Midland Co.

Corporate symbol

Designers: Lester Beall, Inc.:
Clifford Stead, Jr., Lester Beall
Client: Rohm and Haas Co.

Trademark
Designers: Lester Beall, Inc.:
Clifford Stead, Jr., Lester Beall
Client: Caterpillar Tractor Co.

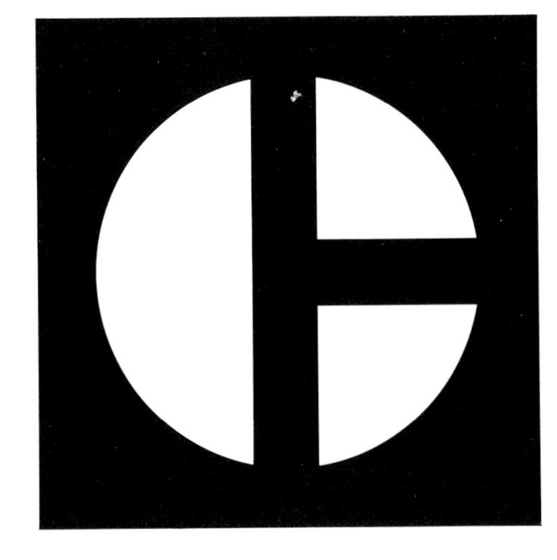

Corporate identity program
Designers: Fogleman Associates, Inc.: W. C. McDade
Client: Interchem Printing Inks

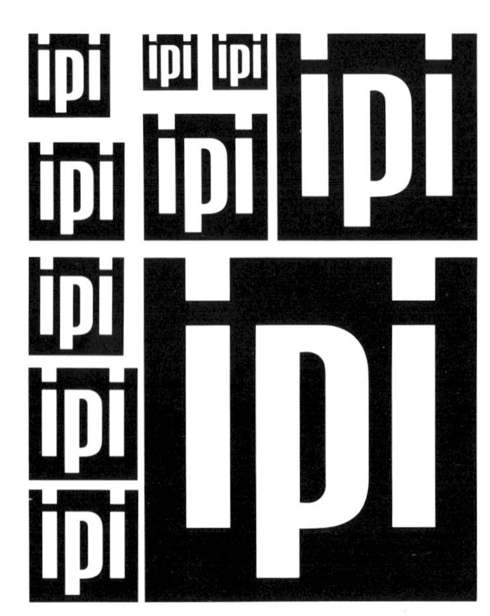

Corporate symbol
Designers: Lester Beall, Inc.: Lester Beall
Client: International Paper Co.

Symbol

Designers: George Nelson & Co., Inc.
Client: Herman Miller Inc.

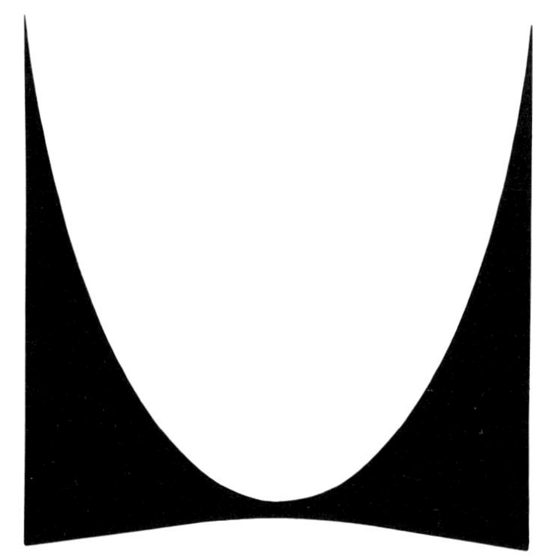

Catalog system

Designers: George Nelson & Co., Inc.
Client: Herman Miller Inc.

Complete furniture catalog can be broken down into specific sales catalogs as required.

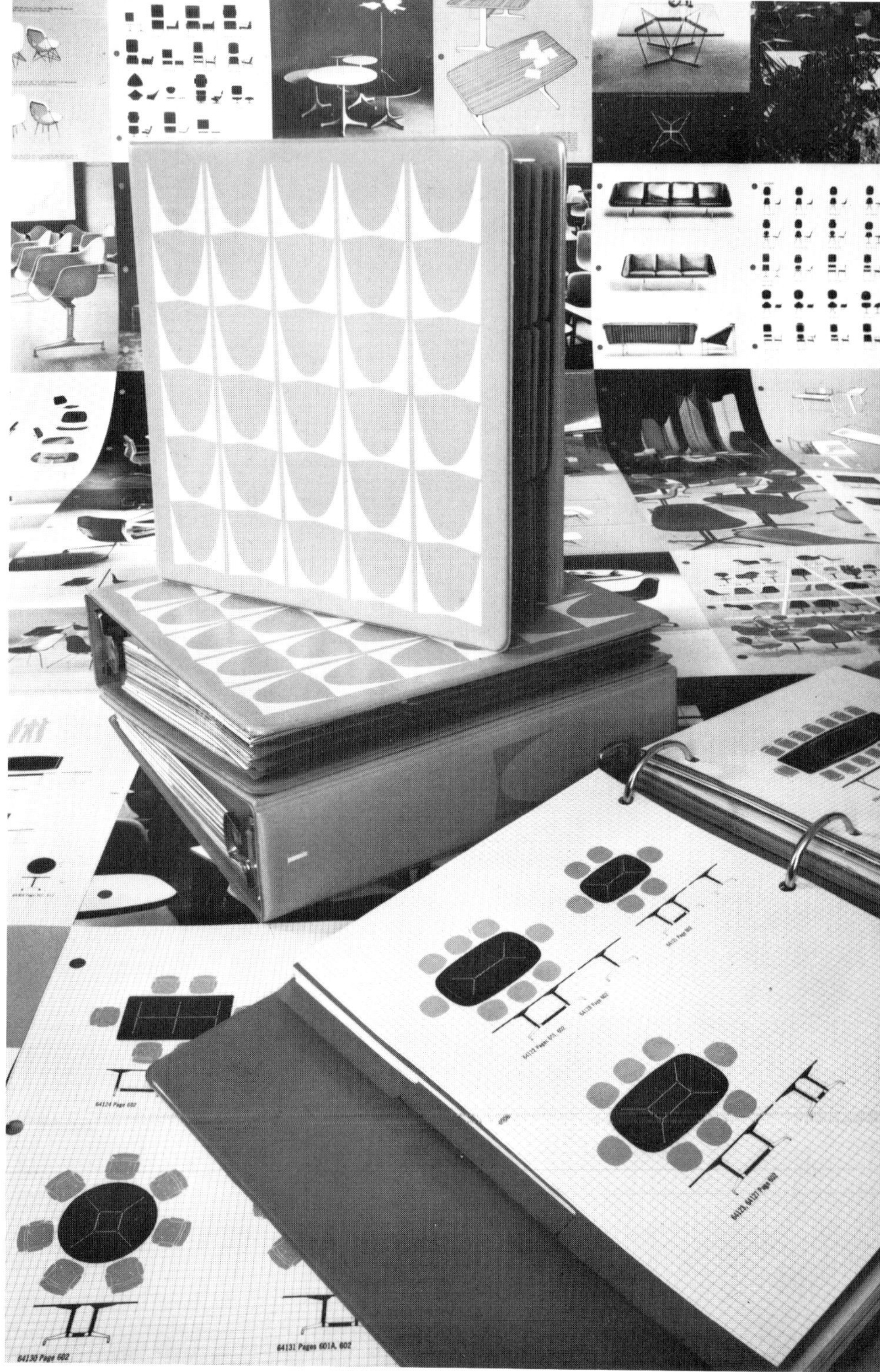

Symbol for film, "Anatomy of Murder"
Designers: Saul Bass & Associates, Inc.
Client: Otto Preminger, Columbia Pictures

Corporate symbol
Designers: Howell Design Corp.: James A. Howell, Madeline Karl
Client: Austin Craig, Inc.

Symbol
Designers: Will Burtin, Cipe Burtin
Client: Lincoln Center for the Performing Arts

Logotype
Designer: Will Burtin
Client: The Upjohn Co.

Logotype

Designers: Chermayeff & Geismar Associates, Inc.
Client: Mobil Oil Co.
Corporate identity program included redesign of all graphics, packaging, signs.

Corporate symbol and logo

Designers: Francis Blod Design Associates, Inc.: Soichi Furuta
Client: Brooke Bond Ltd.

Corporate symbol

Designers: Tepper + Steinhilber Associates, Inc.: Barrie C. McDowell, Gene Tepper
Client: Physics International Co.

Corporate symbol
Designers: Dickens Design Group
Client: Kimberly-Clark Corp.

Corporate symbol
Designers: Lester Beall, Inc.: Lester Beall
Client: Merrill Lynch, Pierce Fenner & Smith, Inc.

Corporate symbol
Designers: Saul Bass & Associates, Inc.
Client: Aluminum Corporation of America

Corporate symbol
Designers: Lester Beall, Inc.: Lester Beall
Client: Connecticut General Life Insurance

Trademark
Designers: Goldsholl & Associates: John Weber, Morton Goldsholl
Client: Foulds Macaroni Co.

Symbol
Designers: Williams Associates: William J. Hannon, Jr.
Client: Information International, Inc.

Symbol
Designers: Raymond Loewy/William Snaith, Inc.
Client: United Nations Development Program

Symbol
Designers: Scherr & McDermott International
Client: Ohio Desk

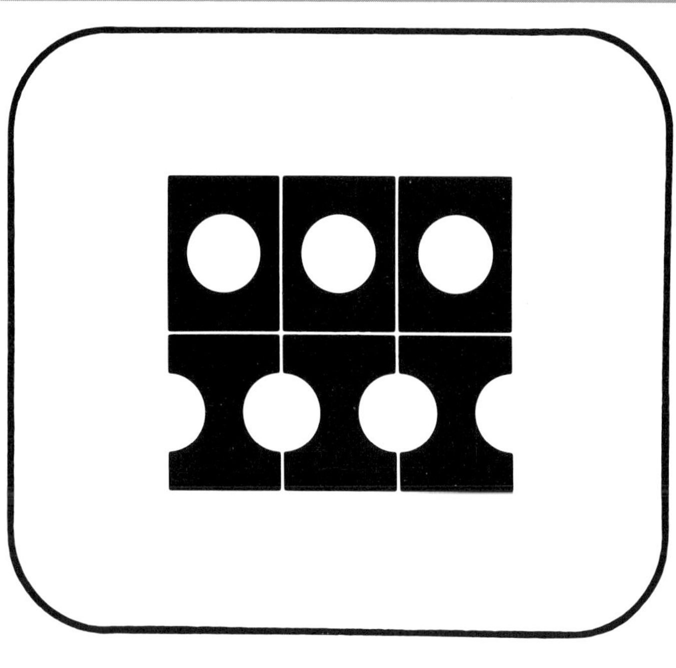

Corporate symbol
Designers: Lester Beall, Inc.:
Clifford Stead, Jr., Lester Beall
Client: MacMillan Bloedel Ltd.

Trademark
Designers: Saul Bass & Associates, Inc.
Client: Lawry's Foods, Inc.

Corporate symbol
Designers: Saul Bass & Associates, Inc.
Client: Celanese Corp.

Corporate symbol
Designers: Dickens Design Group
Client: John Morrell & Co.

Symbol
Designers: Goldsholl Associates: Morton Goldsholl, John Weber
Client: Ben Franklin Stores

Corporate symbol
Designers: Lester Beall, Inc.: Clifford Stead, Jr., Lester Beall
Client: Consolidated Natural Gas Co.

Symbol
Designers: Francis Blod Design Associates, Inc.: Frederic N. Feucht
Client: Savings Banks Association

Corporate symbol
Designers: Becker and Becker Associates, Inc.: Nathaniel Becker, Frederick Leigh, Eugene Grossman
Client: Marine Midland Banks, Inc.

Corporate symbol
Designers: Chermayeff & Geismar Associates, Inc.
Client: The Chase Manhattan Bank

The Package as Product

The function of a package is to contain the product, and there was a time when that was its only function. The packages here have other jobs to do: selling the product, making it more convenient to use, expanding a market, announcing a product change, and even making a new product possible.

These marketing objectives do not necessarily make for beautiful packages, or even tasteful ones, and in the more highly competitive areas of marketing, they tend to have the opposite effect. Successful package design is common, but distinguished packages are rare; probably because package design has to satisfy so many requirements, some of them conflicting. Since the package in the United States is often the chief replacement for the retail clerk, it must describe itself and attract attention. But the package's at-home function is to store and dispense the product efficiently, and to blend into the home environment. Once it is taken home, its attention-getting qualities may become highly undesirable—like an insurance policy that keeps on nagging you to buy insurance.

Package design involves problems of structure or strategy or both. Each package here was judged by the jury to be an effective solution to the particular problems it addressed. Some of the packages were designed along with the products they contain—the Ecko cutlery and kitchen tools on page 163, the Argus line on page 165, the Ber candles on page 162. In other cases (e.g. the A.B. Dick supplies on page 166 or the Schaefer beer can on page 177) the package is one part of an integrated system of corporate identification. The Schaefer design, which was tied in closely to advertising and promotion campaigns, was intended to make beer cans and bottles less kitchen-confined. Based on the discovery that women now buy a lot of the beer consumed in America, the design was thought to appeal to women and to be suitable for appearance in living room and patio.

In personal items, such as toiletries, the question of sex appeal becomes even trickier. Men's shaving supplies ought to look masculine, but must also attract the women who are frequently the purchasers. And some personal products of course are meant to have an epicene appeal, like the VO hair conditioner on page 171.

One of the package's contemporary functions is to play a supporting role on television. Although the commercial is designed to sell the product, it is the package that the consumer usually sees in the store and, in a sense, buys. For a package designer, the question of how the design will show up on television may be as important as how well a logotype will print.

When he designs a complete package—determining the shape of the container and the material of which it is made; determining, and perhaps even inventing, a closure that protects the product in route, and still functions protectively once it has been opened in the home; and devising a surface treatment that will carry all of the desired information (some of it mandatory)—the designer is really designing a product. Like the other products of the designer's craft, packages make up a significant part of what America sees.

Automotive parts packaging

Designers: Ford & Earl Design Associates: Jerry H. Kline, Manuel Jarrin, W. B. Ford II

Three basic design elements: "Picture motif," geometric space break-up, color coding of product lines. Pictures are continuous tone photos converted to line art.

Line of paint cans

Designers: Dave Chapman, Goldsmith & Yamasaki, Inc.: Dave Chapman, William M. Goldsmith, Kim Yamasaki, Eliegey Frasier
Client: Montgomery Ward & Co.

Color-dot concept simultaneously helps differentiate and unite a variety of products.

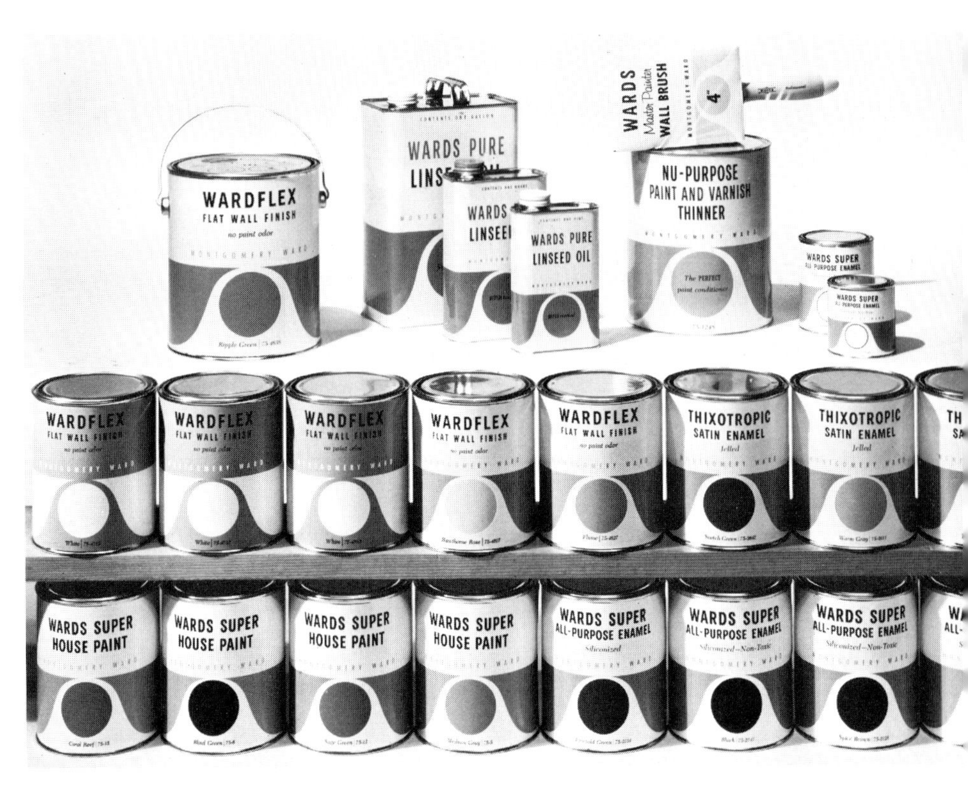

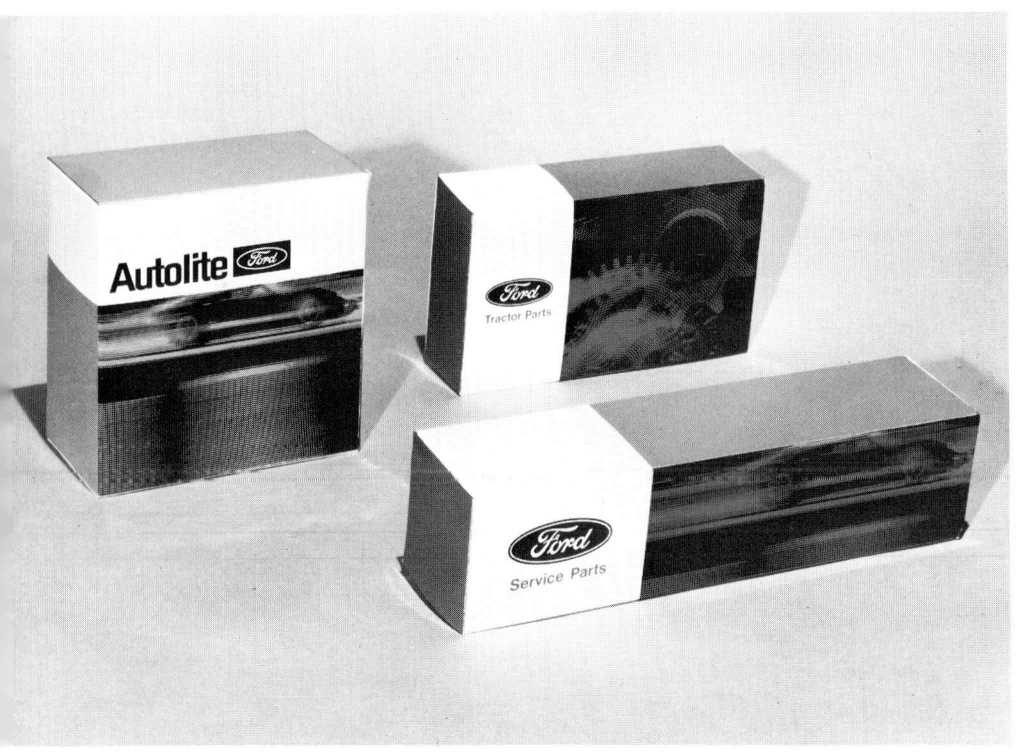

Paint can

Designers: Morton Goldsholl & Associates: Morton Goldsholl
Client: Martin Senour Paints

Stylized spray gun helps distinguish automotive finishes from wall paints.

Product-line packaging

Designers: Lester Beall, Inc.: Lester Beall, Richard Rogers
Client: Stanley Electric Tools

Modular graphic system encompasses wide range of corrugated cartons and boxboard containers.

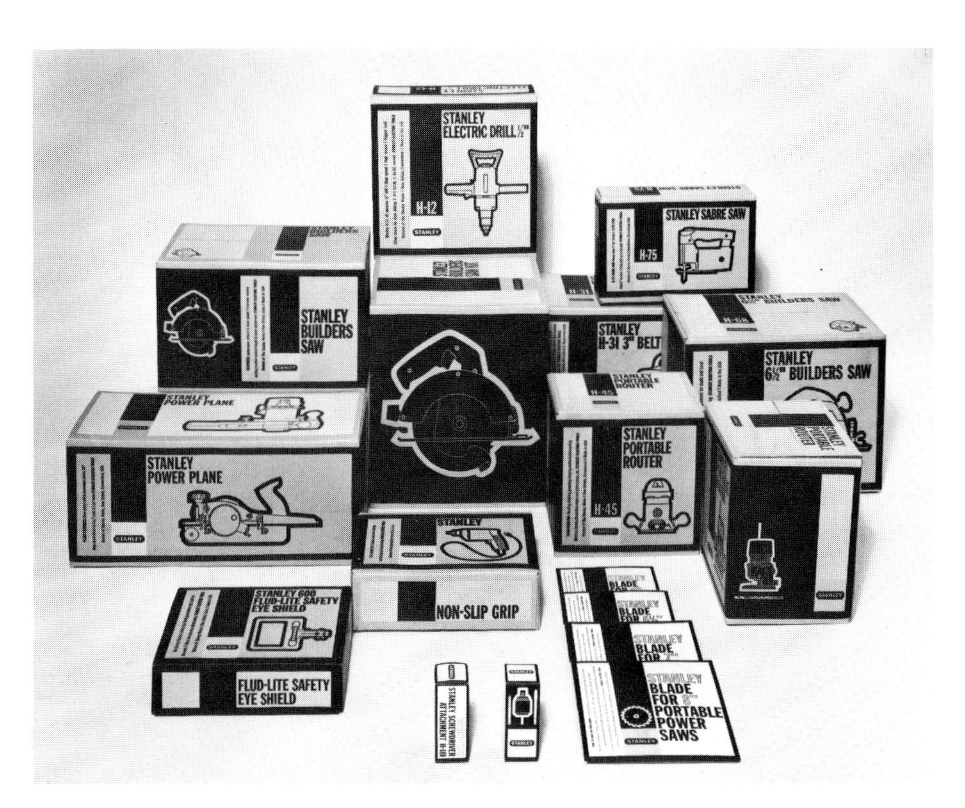

Stacking-candles and display packaging

Designers: Raymond Loewy/William Snaith, Inc.
Client: Ber Design Associates

Match box

Designers: Saul Bass & Associates, Inc.
Client: Ohio Match Co.

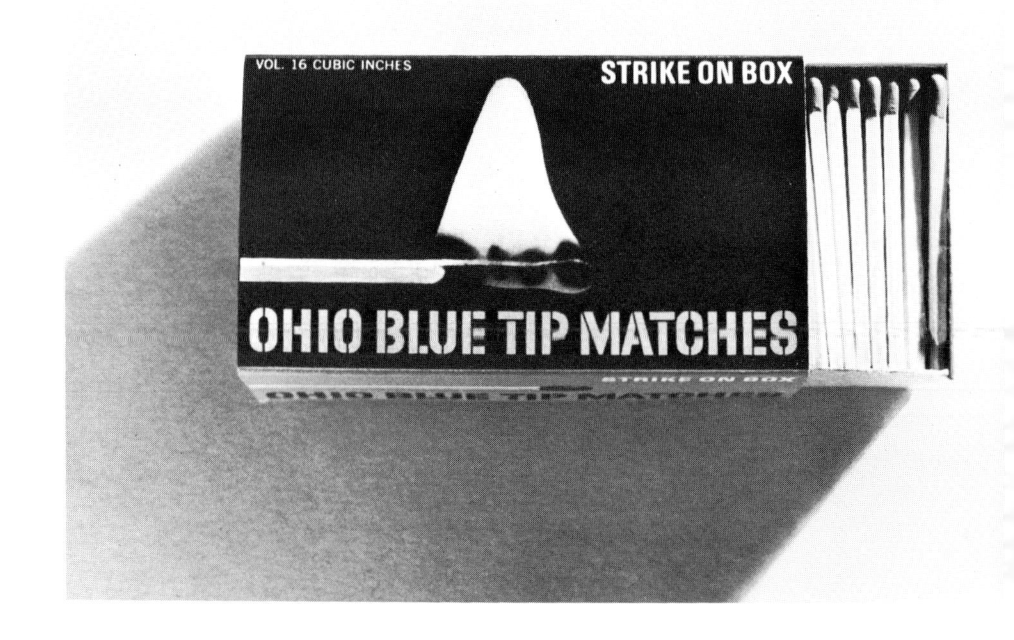

Package
Designers: Dickens Design Group
Client: Polaris Enterprises
White hand holding red and magenta flame signifies special feature of this cooking fuel: It is self-starting.

Cutlery and kitchen tool packaging
Designers: Latham Tyler Jensen, Inc.
Client: Ekco Housewares Co.
Project included design of products, packages, and accessories.

Sponge wrapper
Designers: Dave Chapman, Goldsmith & Yamasaki, Inc.: Hal Hester
Client: Burgess Cellulose Co.
Blue sponge, with white, red and black printing.

Sponge package
Designers: Dave Chapman, Goldsmith & Yamasaki, Inc.: Kim Yamasaki, Dave Chapman, Hal Hester
Client: Burgess Cellulose Co.
Two colors of transparent ink combine with colors of sponges to create color range that could not have been achieved through printing alone.

Package for electronic metronome

Designers: George Nelson & Co.
Client: American Music Sales Co.

Packaging for dictation system accessories

Designers: Yang/Gardner Associates, Inc.:
Peter Quay Yang
Client: Gray Manufacturing Co.

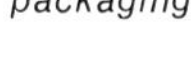

Printed in black and blue, the latter matching the control panel on dictation machine to establish consistency between product and packaging.

Package for stationery

Designers: Goldsholl & Associates:
Morton Goldsholl
Client: Kimberly Clark Corp.

Striped sides relate to other elements in design program; they are, for example, also applied to shipping containers.

Product-line packaging

Designers: Latham Tyler Jensen, Inc.
Client: Argus, Inc.

This graphic house style characterizes all of client's packaging and advertising.

Packaging and labeling for thread

Designers: Alan Berni & Associates, Inc.
Client: Talon, Inc.

Four die-cut windows help retailer check stock. Boxes are color-coded to match spool labels.

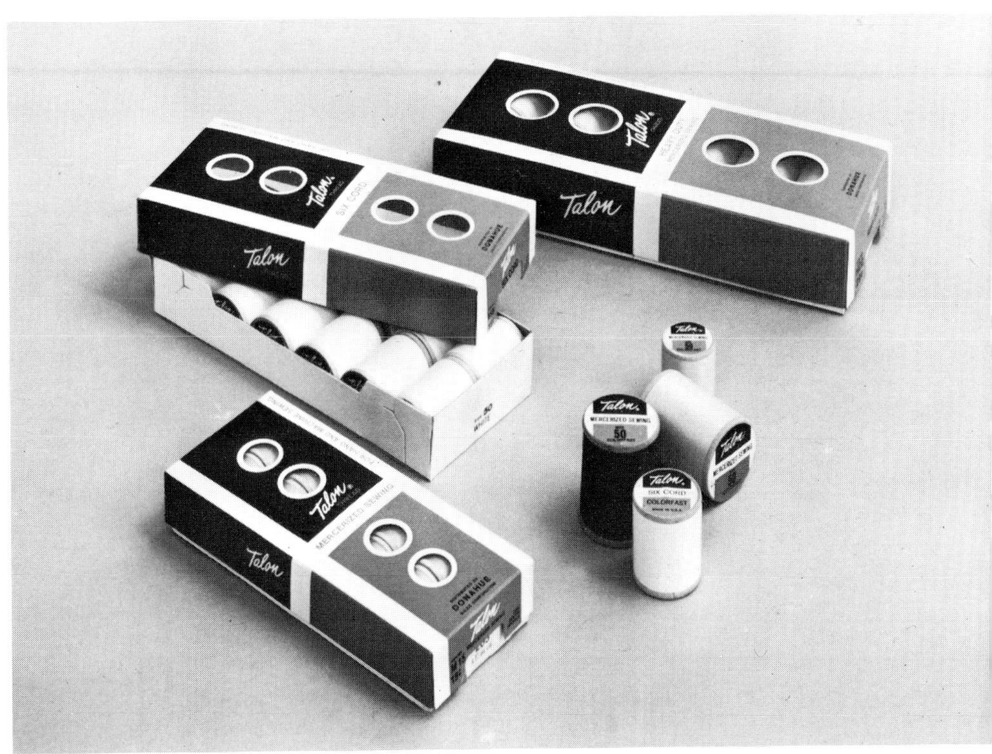

Corporate identity program

Designers: Walter Dorwin Teague Associates, Incorporated
Client: A. B. Dick Co.

Designed to appeal to women, who normally handle these packages and their products.

Packaging

Designers: Brooks Stevens Associates: Brooks Stevens, Wayne Wagner, Ray Anderson, David Nutting
Client: 3M Co.

A three-block module on a white field gives continuity to packaging the client's 28,000 products of various shapes, sizes and markets.

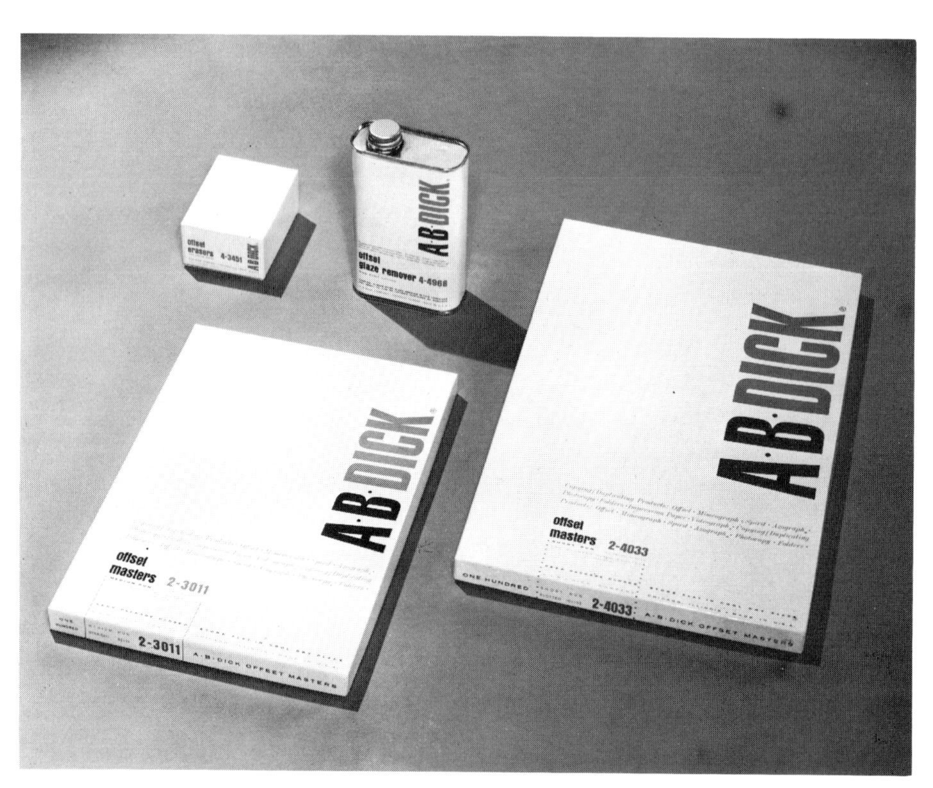

Packages for printing products

Designers: Corning Glass Works staff: Fred Mackie, Gary Solin

Halftone dot design for Corning's graphic media department suggests end-use of the printing product inside.

Packages for thermal wafers

Designers: Corning Glass Works staff: Gary Solin

Graphic design is blow-up of pattern found on surface of the product: a tiny thermal wafer.

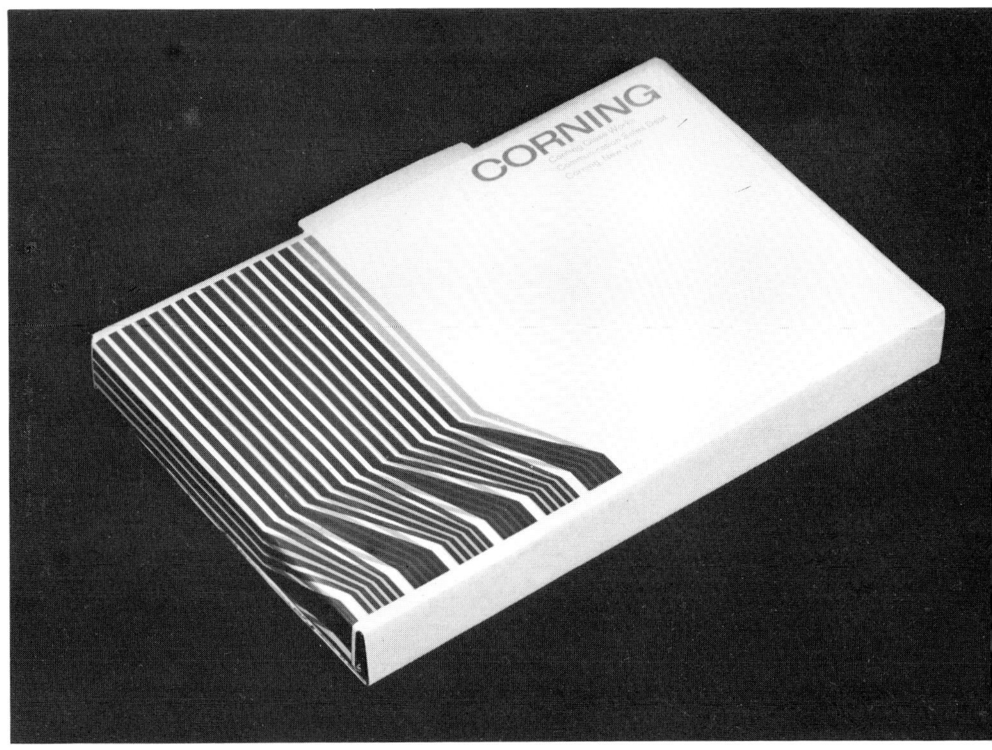

Decorative laminates and sampling system

Designers: Latham Tyler Jensen, Inc.
Client: Textolite, General Electric Co. Chemical & Metallurgical Div., Laminated Products Dept.

Comprehensive design and planning program is represented here by sampling system, in which entire line is displayed in paper folios and "Pop-Packs" injection-molded of linear polyethylene and containing 3" × 3" chips notched to pop on and off integrally molded post.

Package for room-freshener

Designers: Chermayeff & Geismar Associates, Inc.
Client: S. C. Johnson & Son, Inc.

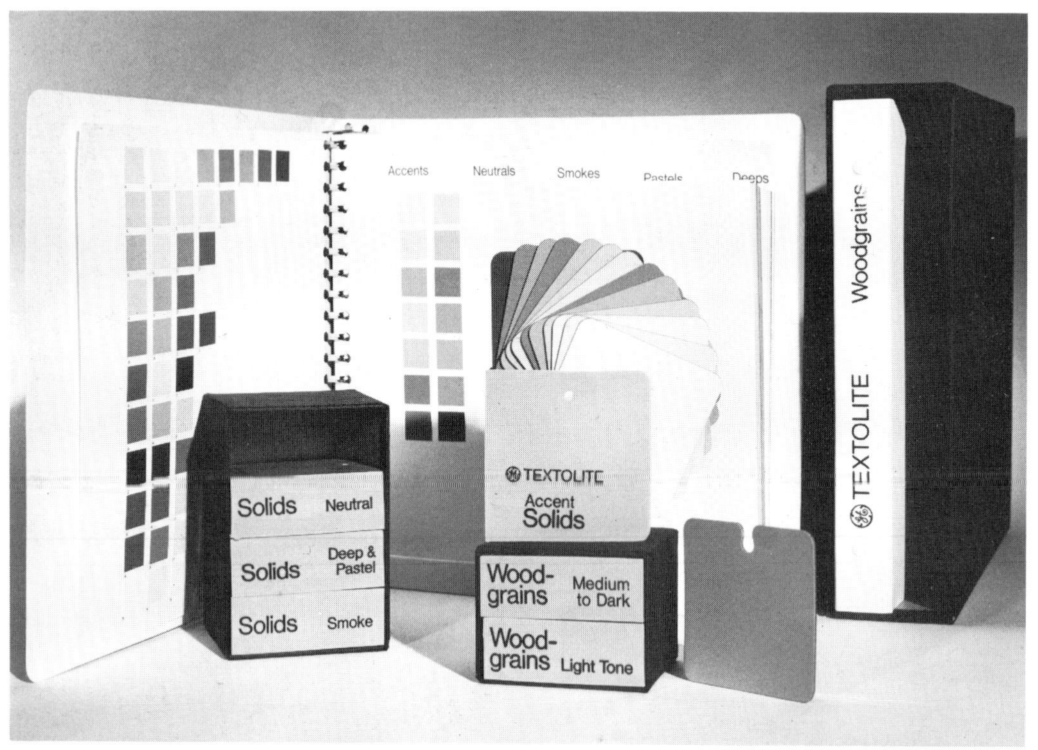

Plastic bottle

Designers: Richardson/Smith, Inc.: David D. Tompkins, Deane W. Richardson. Client's staff: Delmar F. Macaulay
Client: Dow Chemical Co.

Blow-molded polyethylene bottle is easier to carry and pour from than steel cans, thus salable in outlets other than gas stations–for instance, supermarkets.

Plastic bottle

Designers: Dickens Design Group
Client: Climalene Co.

Blow-molded plastic container is white; oval and closure are aqua.

Plastic bottle

Designers: Frank Gianninoto & Associates, Inc.
Client: Thompson Werke GmbH

Trademark is molded into face of container.

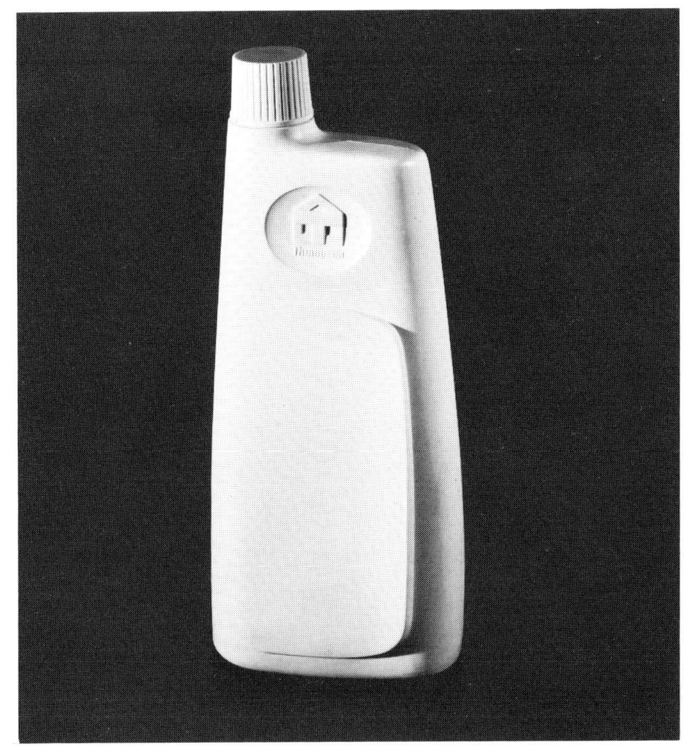

Aerosol shave-cream cans

Designers: Dickens Design Group
Client: Armour Grocery Products Co.

Multi-colored bars surround black Dial logotype.

Deodorant packages

Designers: Dickens Design Group
Client: Armour Grocery Products Co.

White container carries diagonal stripe of metallic blue and green around black logotype.

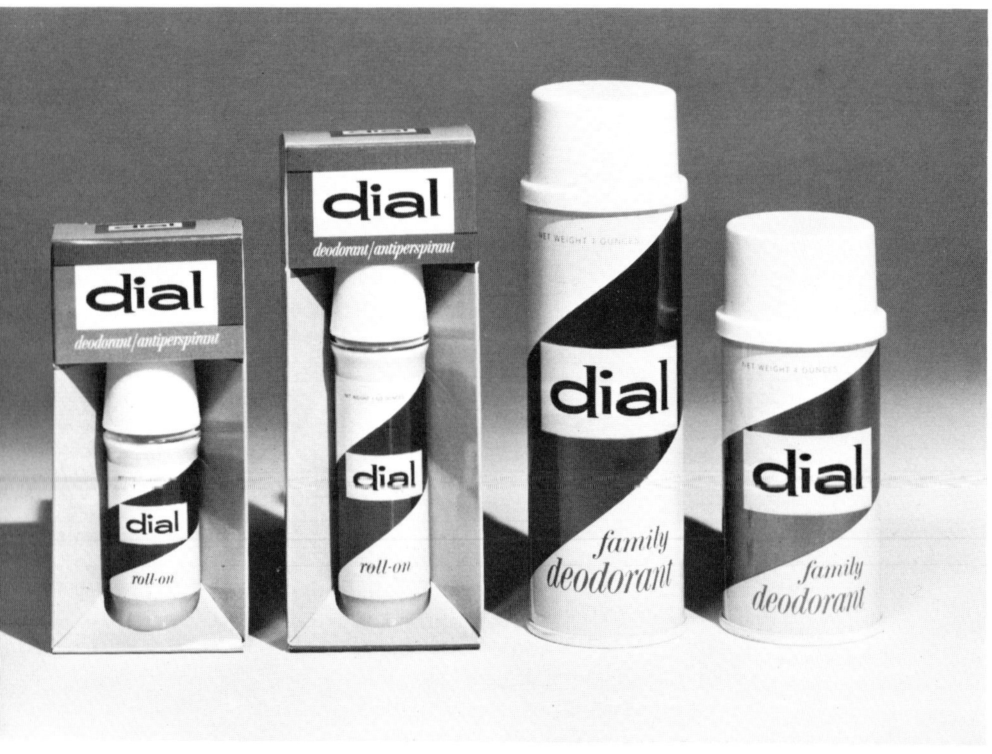

Packaging for line of men's toiletries

Designers: Francis Blod Design Associates, Inc.: George Stehl, Soichi Furuta
Client: The Mennen Co.

Pre-electric shave lotion bottle

Design: Francis Blod Design Associates, Inc.: George Stehl, Frederic N. Feucht
Client: The Mennen Co.

Both bottle shape and closure relate to typical electric shaver forms.

Packaging for line of women's toiletries

Designers: Francis Blod Design Associates, Inc.: George Stehl, Marion Duke
Client: Lehn & Fink Products Corp.

Bottles, graphics and color scheme directed to girls and young women in "middle and upper-lower income brackets."

Hair conditioner package

Designers: Dickens Design Group
Client: Alberto-Culver Co.

Package, like product, is meant to appeal to both sexes.

Deodorant packaging

Designers: Ford and Earl Design Associates: T. Miho, J. M. Earl
Client: Bristol-Myers Co.

Package protects product and displays it.

Perfume packages and bottles

Designers: Raymond Loewy/William Snaith, Inc.
Client: Ondine Parfum, Inc.

Bottle form and closure are adaptable to sizes from 1/8 oz. to 8 oz., and suitable to molding in clear or frosted glasses or to Florentine-finish silver-plated containers.

Mouthwash bottles

Designers: Donald Deskey Associates, Inc.
Client: Johnson & Johnson

Container, closure and label are meant to recall old apothecary jars, and thus project "ethical-drug image."

Mouthwash bottle

Designers: Walter Dorwin Teague Associates, Incorporated
Client: The Procter & Gamble Co.

White plastic closures contrast with green liquid, echo the label color, double as "shot glasses" for individual use.

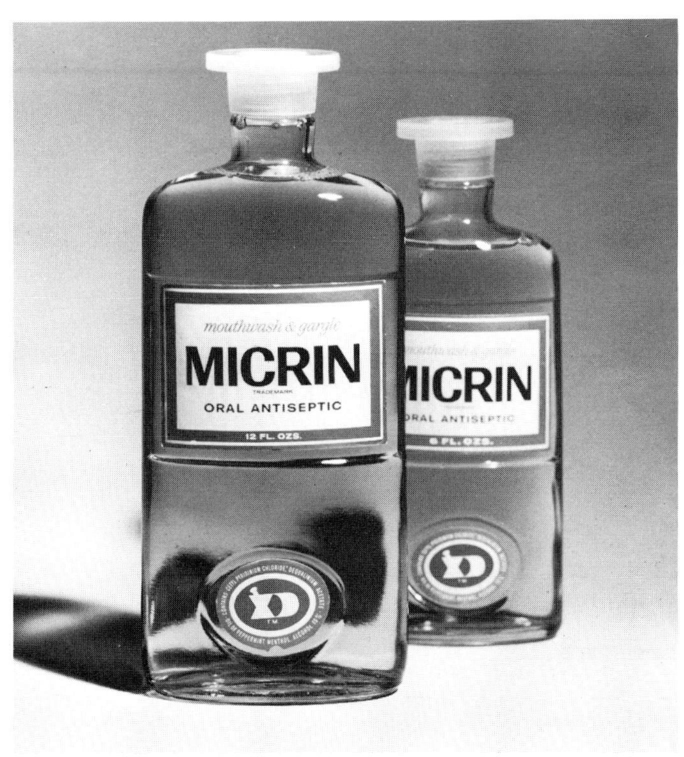

Cigarette packages

Designers: Walter Landor & Associates
Client: Philip Morris Inc.

"Name dropper pak" permits removal of large logo printed on cello outer wrap. More recent Benson & Hedges 100's soft pack echoes wood grain texture using metallic ink in three values of simulated gold.

Sportsman's decanter

Designers: Walter Landor & Associates
Client: Stitzel-Weller Distillery

Molded bottle with blown stopper is illuminated with sportsman's symbols.

Liqueur bottle, graphics, packaging

Designers: Ford & Earl Design Associates: Manuel Jarrin, Pamela Waters
Client: George M. Tiddy & Sons Ltd.

Concepts of tradition and quality are expressed through such devices as antique gold foil and bilingual labeling, while Optima lettering connotes modernity.

Water bottle

Designers: Walter Landor & Associates
Client: Arrowhead Puritas Co.

Design objective is candidly stated: "to enable the packager to gain a foothold in a market where there is no product difference." Form makes it possible to pour from this ½ gal. bottle without lifting it.

Beer bottle, beer can

Designers: Walter Dorwin Teague Associates, Incorporated
Client: F. & M. Schaefer Brewing Co.

Design is also used in advertising and promotional material.

Bottle and carrier-carton

Designers: Morton Goldsholl Design Associates, Inc.
Client: Seven-Up Co.

Pinwheel symbol used doubly on bottles and in single billboard style on multiple carriers. Carriers are printed black, with one accent color; upper logo of bottles is white, while colors of lower logo and pinwheel alternate–making possible a multi-color display from stock clerk's random stacking.

Jelly and jam jars
Designers: Dickens Design Group
Client: Mr. Polaner & Son, Inc.
Urnlike shape enhances display of product and ease of getting it out of the jar. Large P is color coded to contents.

Packaging program
Designers: Lester Beall, Inc.:
Clifford Stead, Jr., Lester Beall
Client: House of Herbs, Inc.
Program includes various sizes and varieties of bottles and labels, shipping and gift cartons, and modular plastic wall racks (shown here).

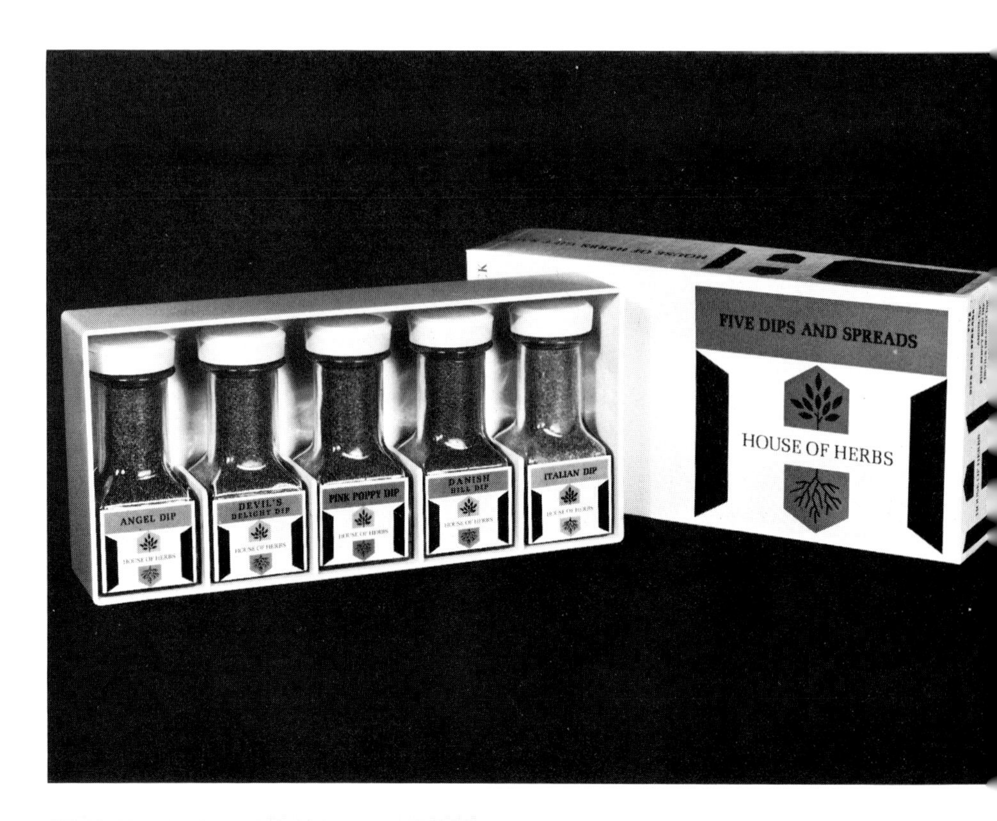

Candy packages

Designers: Ford & Earl Design Associates: Pamela Waters
Client: The J. L. Hudson Co.

Bright colors and casual type faces are used to promote inexpensive party candies.

Cookie packages

Designers: Raymond Loewy/William Snaith, Inc.
Client: National Biscuit Co.

Paper bags are intended to convey a "pastry shop image."

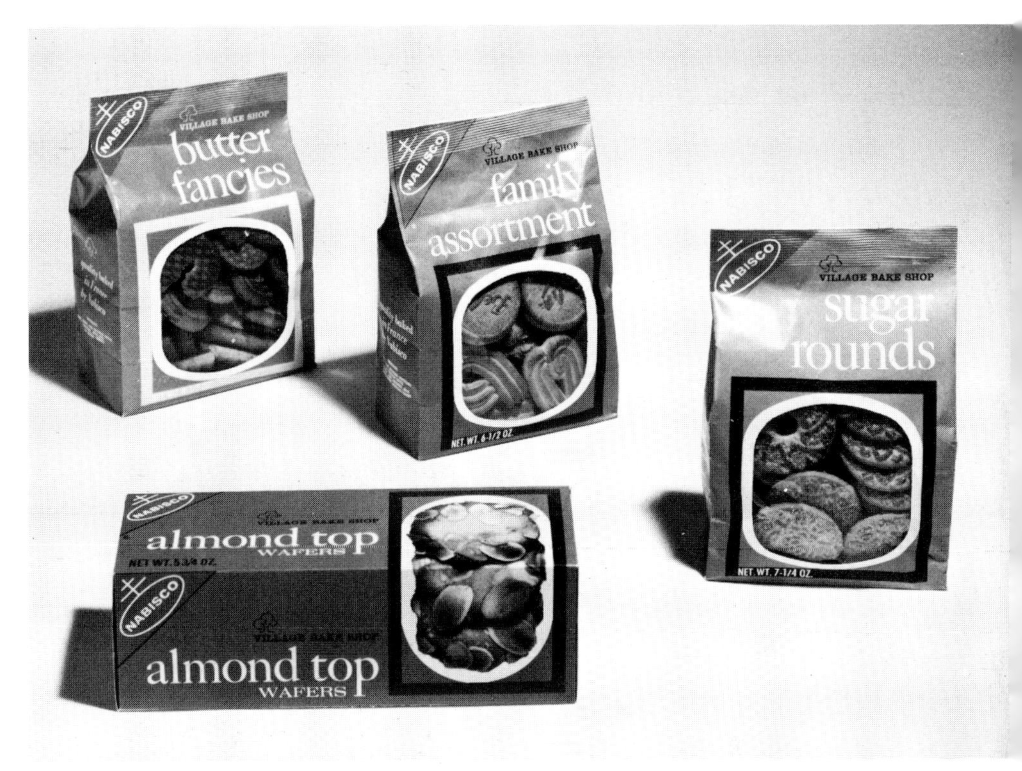

Credits

IDSA Editorial Board

Nathaniel Becker, Chairman
Arthur BecVar
Dave Chapman
Henry Dreyfuss
Charles Eames
George Nelson
Arthur Pulos
John Vassos

IDSA Jurors

Product Design
Eugene Bordinat
Donald McFarland
Joseph Parriott

Communication Design
Lester Beall
Francis Blod
Morton Goldsholl

Environmental Planning and Design
Walter B. Ford, II
Robert J. Harper
Jens Risom

Coordinating and Review
Jay Doblin
Richard Latham
Eliot Noyes

Editorial Direction

Ralph Caplan

Design

Bert Waggott

Coordination

Mary Belden Brown

IDSA Board of Directors

Robert Hose, Chairman of the Board
Tucker Madawick, President
William Goldsmith, Executive Vice President
Arthur Crapsey, Vice President
Richard Hollerith, Vice President
Arthur Pulos, Vice President
Eugene Smith, Vice President
Peter Quay Yang, Secretary
Olle Haggstrom, Treasurer

James Balmer
Nathaniel Becker
Eugene Bordinat
Dave Chapman
Niels Diffrient
William Hall
Paul Karlen
Richard Latham
Don McFarland
George Nelson
Joseph Parriott
John Vassos
Read Viemeister
James Alexander
John Christian
Hin Bredendieck
Pierre Crease
George Kosmak
Robert Redmann
Benjamin Werremeyer

Index

E

F

G

H

U

V

W

X

Y

Z

CONCORD, MASS.
CONCORD
MA
PUBLIC LIBRARY